Bear River

Waterfall on Ostler Fork
of the Bear River.

Bear River

Last Chance to Change Course

Craig Denton

Utah State University Press
Logan, Utah

Utah State University Press
Logan, UT 84322-7800

Publication of this book was supported by a subvention from the
Willard L. Eccles Charitable Foundation

ISBN-13: 978-0-87421-650-9
ISBN-13: 978-0-87421-664-6 (e-book)
Manufactured in China

Library of Congress Cataloging-in-Publication Data

Denton, Craig, 1947-
Bear River : last chance to change course / Craig Denton.
p. cm.
Includes bibliographical references and index.
ISBN 978-0-87421-650-9 (cloth : alk. paper)
1. Bear River (Utah-Idaho)--Pictorial works. 2. Bear River (Utah-Idaho)--History. 3. Natural history--Bear River (Utah-Idaho) 4. Bear River (Utah-Idaho)--Geography.
5. Bear River Valley (Utah-Idaho)--Biography. 6. Interviews--Bear River Valley (Utah-Idaho) 7. Bear River (Utah-Idaho)--Environmental conditions. 8. Bear River
Watershed (Utah-Idaho)--Water rights. 9. Water use--Bear River Watershed (Utah-Idaho) I. Title.
F832.B3D46 2007
979.2'13--dc22

2007008036

To Stacie and Gavin, for enriching our family.

The wintry Bear River below Oneida Dam.

Contents

Preface

When the eminent documentarian Sebastião Salgado was asked where he gets his ideas, he replied that he discovers the seeds of a new documentary in the one on which he is working. Such has been the evolution of this documentary on the Bear River.

My earlier book, *People of the West Desert,* documented the lifestyles, hopes, and dreams of people living in the high desert of western Utah and eastern Nevada. In writing it, I learned how fundamental water is in their lives. So, about five years ago, when schemes to dam the Bear River and move the water to the Wasatch Front began to crop up in news stories, my antennae went up. I already had realized that water will be the primary political, social, and economic issue in the Intermountain West in the twenty-first century. Looking at how the river that carries that precious resource to its stakeholders is being used seemed to be a good way to talk about these critical issues.

Being a fisher, I've always been attracted to rivers anyway, and like all fishers, I dream of finding "new" water, perhaps some undiscovered spot where I can cast a fly before the word gets out. Although I knew that the Bear wasn't known for its fishing, I told myself, you never can be too sure. Some clever fisher might have started that fable to protect his or her favorite hole. So exploring the Bear seemed to be a good way to satisfy my river lust while also learning about a river that has played such a dramatic role in the development of northern Utah, western Wyoming, and southern Idaho.

Planning a documentary, especially one that is such a "big" story—geographically spread out and complicated with knotted, emotional issues—takes some time. This documentary was supported in its initial stages by Faculty Fellow and sabbatical leaves from the University of Utah and a development grant from the College of Humanities.

I spent the first year of the project doing fieldwork to get a general feel for the river, and then the next six months in various libraries working my way through scientific papers and books that talk about the geology, hydrology, geomorphology, ecology, and history of the Bear River and its basin. That subject matter is woven into the first four chapters of this book. Then I spent weeks in the field in all seasons, trying to capture the features of the river and its surrounding environment photographically to provide the reader with the most accurate picture of the Bear possible in a necessarily limited amount of space. To get to some places, I backpacked. Other times, I paddled a canoe. But most of the vantage points are accessible by vehicle and short walks. In some instances, landowners graciously gave me permission to photograph from their property. Most times, however, the images were taken from viewpoints on public land. Sometimes, when I wanted to take a shot from a pivotal position, I decided against it because I didn't know who owned the land. In just a few cases, it was pointed out that what I thought was public land was not, and I appreciate the tolerance those landowners gave me. One man with a Napoleonic complex couldn't be swayed. Ownership of the Bear and its adjacent lands tends to be a complicated and fluid issue, as well as a philosophical problem.

I photographed with a four-by-five-format field camera and a 35 mm camera. I used the first to capture important geographic features in the greatest detail possible, and the second for documenting the activities of people, who have a tendency to move faster than bulky photographic equipment can follow. I shot on both color and black-and-white film. The images in this book come from that black-and-white portfolio. We've been able to print them as double-dot black duotones thanks to a grant from the Willard L. Eccles Charitable Foundation, for which I'm eternally grateful. The color images are the crux of the accompanying photographic exhibit.

The book would have been incomplete, however, without including the human side of the story. While I was doing initial research, several names cropped up repeatedly, people who have a pivotal role in the way the Bear River is used. Or I stumbled across people who represent the interests of groups who also have had a long relationship with the river, even if their names haven't made the news. I interviewed many of these people, and their voices and perspectives are woven into this book. When I informed them about my project and my goal of a book and an exhibit, they graciously consented to be interviewed. I taped the interviews, with their permissions, to retain the accuracy and authenticity of their voices. How these interviews have been integrated into the discourse is solely my responsibility.

Yet my primary concern from day one in the evolution of this documentary was to discover and preserve the voice of the Bear River. Perhaps that's due to my long attachment to western rivers. I've also been troubled by how little attention human beings today give to our natural systems. While concern for the environment was one of the dominant themes of the 1970s, people seem to have forgotten how we learned why it is so important. A healthy ecology and environment have fallen several rungs on the ladder of ideological hierarchy. We seem to be in an era of gross hubris: demanding consumption before understanding, and thinking that we are in control of our lives and can fix whatever needs fixing if it really gets bad. Or the corollary might be true: we think we cannot control anything, so it's every person or economic interest for him or herself. I've tried to shine a light on the landscape between these two destructive views.

I'm also concerned that not all stakeholders in the Bear River have been able to make their voices heard. The plants and animals of the basin always will be largely secretive and silent. I've tried to provide an occasional glimpse of them, but they remain in the background. There really should be an entire book that speaks for them and tells their stories. Regrettably, there wasn't enough time and space to make them come fully alive in this documentary. Moreover, it's difficult to avoid anthropocentrism when writing a book produced by and for human beings.

The problem is one of language, or more precisely, what language to use. Probably using words to give voice to a river will ultimately fail because words are uniquely human symbols. Words are loaded with denotations and connotations at the same time. Words are shaded with political power. Words, and the verbal language systems that establish their shared meanings, tend to create boundaries and binary representations. By defining what is, a word also suggests what is not. Inclusion inevitably means exclusion, and what is outside the province of the human symbol becomes "the other," whether it's a plant, animal, or spirit, like the voice of a river.

The notion of voice is tied to an oral tradition of language. Voice, too, is the province of aural delivery and auditory perception. Occasionally, a river has a voice that literally can be heard, but the voice of a river is much more than that. Consider this scenario as an illustration of the difficulty of using words to give voice to a river. If a river crosses an international boundary, and each of the nations uses a different language, the words used to give voice to that river would change at the boundary, but the true voice of the river would not. Its voice transcends language.

At most, words and language can become the voice of a river through metaphor and personification. For that reason, I've tried to capture the major tones of the Bear River's voice in interludes that precede each chapter.

I ask the "reader" of this book to think of photographs as another language, one that may be better suited to give voice to the Bear River. The visual voice doesn't draw boundaries. It doesn't create a speaker and separate that person from the other, outside world. The visual voice is fluid and integrative.

Visuals don't need a metaphoric equivalent. They are not *like* or *as*. Visual images *are*. For that reason, they don't need codification or translation. They don't need a dictionary or historical use to establish meaning.

Granted, photographs do create a spatial boundary: the frame of the camera's viewfinder. Moreover, that framing eye is not the eye of the river. If there is such a thing as a river's seeing eye, it is molecular and outside the ability of human technology to record as light. The photographs in this book have been captured through a particular human filter. I realized that, and, as much as possible, I tried to shoot the photographs at the river's level, feeling that was the perspective closest to the river's point of view. Although sometimes I shot from a higher camera angle to provide a greater sense of place, to show the Bear's meandering and how it interacts with the land, aerial photography possesses an inherent problem. Aerials have a detached quality. They are less tied to the earth, unlike a river.

No doubt the concept of a visual voice is jarring and seems illogical. That's because our logical thought processes are the domain of words. So, if you can't quite accept the visual as the voice of a river, think of the visual voice as complementing and magnifying the verbal one. Consider the notion that the visual voice can extend the range and pitch of the verbal one. If the photographs in this book can't become the singular voice of the Bear River, they may still be its most effective translator.

Documenting the voice of the river, then, hasn't been easy, and sometimes I've wondered whether trying to do it is a quixotic venture. I've had to ask myself, how can I really speak for the Bear, either in words or through photographs, when the human voices surrounding the river are so insistent and powerfully entrenched—and when I am part of that human enterprise and benefit from it. Even when I first hoped I could focus the project's attention solely on the voice of the river, I soon realized the result likely would be shallow and precious. Human stakeholders are a necessary and important part of the ecosystem of the Bear. So I've tried to blend their voices with that of this river and all rivers. This book is a choir with multiple stakeholders taking different parts, sometimes singing in harmony, sometimes in discord. However, I trust the reader can still hear the soloist—the Bear River.

Acknowledgments

Acknowledging people who have helped a book come together is always a humbling experience because reflecting on these contributions drives home the point that any creation is a collaboration. This book is no different.

I am deeply indebted to the people I interviewed who have been woven into the text of this book. Their insights and gracious willingness to share them have enriched this documentary. They've contributed immensely to whatever illumination this book provides.

There are other people, too, who have helped along the way but aren't identified in the text. I'd like to thank them here: Zach Frankel, former executive director of the Utah Rivers Council gave me background and source information on problems facing the Bear River. Fred Selman shared his research on the sometimes-murky locations of fords and ferries on the river. Val Grant, former president of Bridgerland Audubon Society and head of BioResources, Inc., offered his insights on the river in Cache County, Utah. Vince Lamarra and Hart Evans of Ecosystems Research Institute documented the water-quality measurements in various reaches of the Bear. Jack Schmidt, professor in the Department of Watershed Sciences at Utah State University, provided an overview of the Bear River Basin. Richard Toth, professor in the Department of Environment and Society at Utah State University, shared his graduate students' study of the geomorphologic, ecological, and hydrological dimensions of the Bear River. Dottie Kasperson of Preston, Idaho; Kent McMurdie of Deweyville, Utah; and Gordon Zilles of Hyrum, Utah, allowed me access to their land.

Blaine Newman, recreation specialist at the Pocatello office of the Bureau of Land Management, provided me with reports on Idaho's appraisal of possible Wild and Scenic status for the Bear River. Kerry Brinkerhoff of the Friends of the Native Americans of Northern Utah showed me important sites in the history of the Northwestern Band of the Shoshone Nation. Paul Knopf from the Department of Planning and Development in Evanston, Wyoming, responded graciously to my questions regarding the town's river-reclamation efforts. Dan Miller of the Bear River Watershed Council

Coniferous pines and firs in the subalpine zone of the Bear River headwaters.

helped unravel the problems of watershed protection in the Bear River Range. Phil Donegan of PacifiCorp worked through the corporate bureaucracy during a time of heightened security to give me access to the Grace power plant. Monte Garrett of PacifiCorp gave me the record of nongovernmental-organization claims during the Grace relicensing process.

And, to the two outside reviewers of the manuscript, your critiques highlighted issues I had missed. Your suggestions have helped create a better book.

A reflective moment at the inlet of Allsop Lake.

Gravity my engine
Water my soul.
I am the teller of life and deep time.

You would measure me,
Sever me,
Own me,
In your name.

Let me flow
In your imagination
That I may speak.

Allsop Lake is the source of Left Hand Fork of the Bear River and the most easterly lake in the river's watershed.

Defining the River

It's hard to get hold of a river. It invites the touch, but it's difficult to grasp, an elusive thing that exists as much in the imagination as on the ground.

Most times, a river knows its place, sticking to hollows it carves for itself in the earth. In high times, though, it wanders where it wants, with blind momentum and its own cadence.

A river's personality changes from day to day, sometimes shyly, sometimes with braggadocio. Some seasons, a river can be secretive and timid with a flow that struggles to cover its bed, eventually drying up on sun-baked rocks. Other times, when provoked, it can fill with thunder and fury.

The definition of a river seems simple enough. It is flowing surface water of a size large enough to capture the imagination and have a name. Rivers share that ineffable magic of the way water forms and holds together, with hydrogen and oxygen atoms sharing each other's electrons so they can become whole molecules. Then the molecules, which look like Mickey Mouse heads, link themselves together through the mutual attraction of two positively charged hydrogen ears bonding with an adjacent, negatively charged oxygen head to create a chain that has mass but no form.

Somehow that definition misses its mark. To paraphrase Dylan Thomas, water may be composed of two parts hydrogen and one part oxygen, but there is something else. And no one knows what that is. A scientific description can't explain the eloquence of a river or its complexity. The fact that a river is water only tells part of its story.

A river is a child of gravity and wholly dependent upon it. It's possible for water to climb uphill a short way via capillary attraction, but that momentary wandering is a fanciful conceit and not the unyielding path that a river must follow. Gravity is a stern arbiter. It allows only one way—down—and a river must take that route. A river bends in concert with the contours and opportunities of the land, but it is single minded in its goal.

It's instructive to think of a river as a medium. As gravity provokes it to move, it bears life's debris in its currents. Because of a river's intimate contact with land and its naturally erosive action, it carries geological sediments, those inorganic minerals that are the building blocks of new life. More importantly, a river carries with it the history of the land in its basin, as well as a deep time narrative of the earth: rocks formed, then deformed, then formed again eons ago. They slowly move from a river's source to its mouth, sometimes providing a cobbled bed for the stream, other times littering the floodplain.

A river is life's elixir, too. Along with minerals, a river carries organisms: bacteria and single-celled plants and animals that become the nutrients for successively higher-order animals. Moving water is a delivery system and incubator for aquatic life on land. It plenishes a community that extends inland beyond its banks to form a lotic or riparian ecology.

A river is life's refuse collector. Besides living organisms, it bears dead organic material. Leaves, twigs, and branches falling into or carried along a stream provide other nutrients for the lotic system. Plants drop seeds into the water to spread their life. Plants and animals evacuate their wastes into a river and make it fecund. The highest-order animal, *homo sapiens,* often uses a river to hide its garbage or, at least, move it downstream to the point where it isn't a problem for the local community.

Sometimes a river is not quite itself. Instead, we see it as a reflection of other beings or other moods. Then it is more lyrical than practical, more image than object. It can be a brilliant, specular reflection of the sun early or late in the day, a glaring ribbon that burns the retina. Other times, a river is an ethereal, deep blue as it reflects the heavens. If we move closer and choose a different angle, a river reflects life that crowds against its banks: willows, cottonwoods, or a deer slipping to its edge to drink under the cover of dusk. Sometimes, when dense fog forms above rivers where warmer water meets colder air, a river seems to disappear, blending into a gray union with its sibling.

Looking straight down sometimes the water is clear, and the river reveals its innards. Then, when snows melt in spring or late summer storms bring cloudbursts and flooding, or when a river simply moves over remnant silts left by ancient lakes, the water clouds, and the stream becomes mysterious and threatening.

These are the musings of the poet's sense of a river. Other human beings, however, have their own definitions, usually more prosaic and utilitarian. More importantly, those definitions are laden with values and assumptions about the river's place in a human cosmos.

For centuries, human beings have tried to overlay rationalism on the nature of rivers. As scientists, we record annual flows, breaking down the discharge rates according to seasons. We carefully measure the high and low water marks and the potential lateral range of a river when it overflows its banks. We dutifully list average temperatures in segmented reaches and note the dissolved oxygen count. We catalogue turbidity. We analyze what a river carries: its biomass, inorganic compounds, total dissolved solids, pollutants. We seek to understand compositions and behaviors of rivers, usually because the relationships between a river's deposits and its unique currents become markers of that river's wealth—in human terms.

When the Bear turns south at Soda Point near its eponymous dam, the river continues to cut through relatively young volcanic rock, carving the walls of upper Black Canyon.

Gathering longitudinal data on rivers also gives us potential insight into ways of controlling them. Data make rivers more concrete by rendering them understandable, and since the data are measured uniformly from river to river, our measurements can make rivers more unique because they can be classified and ranked according to those figures. Knowledge about rivers invites comparisons. Some are better for boating, others for fishing, and public dollars can be directed to those ends.

The more we know about rivers, the more predictable their behaviors become. Predictability provides opportunities for cost-effective use of a river. For instance, dam building, the sad end state of a river preferred by some, became politically feasible when river storage data became available. The trouble is that those data often were massaged or misinterpreted to support foregone conclusions.

Once we can quantify a river, we can parcel it out, leading to the incongruous situation that, although a river is fluid and continuous, data provide a way to fractionate it. Claims can be made on those parcels. But, because there are always more claims and rationalized needs for water than can be met by nature's finite resources, conflict over water is inevitable and incessant.

Human beings are the only ones of all God's creatures who try to get hold of a river by claiming ownership. Water law is a human construction that attempts to create order from the chaos that would reign if a river were free of regulation. Water law provides a way to resolve human conflicts, often at the expense of other species. Water law offers a mechanism for political divisions—national, state, and local—to negotiate use of river water. The assumption of human dominion over rivers becomes visible in the names we give to those who are responsible for controlling rivers: water master, river commissioner, reclamation manager. Water law also tacitly acknowledges the damage it creates by providing avenues for riverine mitigation.

Our current water laws in the West are the residue of the need for frontier miners to have a reliable flow of water in an arid region. The effect has been to create a Byzantine patchwork of rules and regulations that rely more on precedent than logic. To work their claims, miners needed to be able to divert water. The right to divert evolved into a system that rewarded those who were there first. Now water law neatly ranks claims to river water based on seniority. "First in time" mandates the division of water rather than "first in need," and senior water rights must be served before junior ones, even if that means that these rights holders get no water during a drought year.

Water law also evolved around the dictum that river water must be used and, to be used, must be diverted. This produced the catch-22 for the river. Water had to be diverted to create a beneficial use, and leaving water in a river was defined by law as nonbeneficial. So, according to water law, a river is only a conveyance of water rights and exists by permission once humans measure and divvy it up.

Water evolved as a property right, and hence, a river is something to be owned. Some state laws say that property ownership ends at a river's edge. Others maintain that it extends to the riverbed but allow trespass rights to those walking along the bank between the river's edge and its high-water mark. Still other states say that no one can have access to a river without the streambed owner's permission. While the federal government reserves rights that predate all private claims, states maintain that they have ultimate authority over the way water is used within their boundaries.

When a river became property, it became something to be coveted or hoarded—something to war over. Even now a river is subject to public demands and legislation for expensive, Rube Goldberg-like projects that try to move it to another place, another basin, or even uphill. Most would say that one cardinal principle defines a river: it flows downhill. But when water law evolves to the point where a river may even run uphill, somehow we've lost the sense of a river. We've lost track of its voice.

Or perhaps we've co-opted it. By defining rivers according to human yardsticks, we've substituted our loud, demanding voices for the more opaque voice of a river. Even when we try to imagine the essence of a river, we are forced to use metaphor and personification, those anthropomorphic figures of human speech. There is irony and frustration in any attempt to define a river. Inevitably, we must write with language that has evolved from human experience. It may not be the language a river uses, but it's the best we've got.

Given all this, can a river exist unto itself? Can a river have its own story, or is that story necessarily about humans because we write it? Can a river have a spirit or presence, or is that a human fancy that winds up sounding effete when attributed to a river? Can a river have a voice, even if it has to be interpreted?

Many would say that a river must be given a voice. Only then can it tell its stories and reveal its different moods. Equally importantly, a river must have a voice if it is to gain access to political debate, to legal standing—to power. Only when a river has a voice can it become a player in its own grand morality play, where verbally armed camps fight over its soul and substance.

Ironically, if a river were given a voice, human interests might become the primary benefactor of its cry. No longer would a river be seen only as a conveyer of water rights. If a river voiced its stress—and humans could learn to truly listen—a river might tell us what we need to hear.

This, then, is the story of the Bear River and the record of its voice. In some ways, the Bear is no different from other rivers. It has multiple personalities. Its moods change over distance and time, dramatically or almost imperceptibly. It dons different guises, sometimes solid, sometimes vaporous, but mostly liquid as it slips through stunning mountain bowls, through fertile valleys to a saline sink. Such is the magic of rivers and water.

Most people experience the Bear River at freeway speed. It slips below them as they cross it on Interstate 80 at Evanston, Wyoming, or Interstate 15 near Honeyville, Utah. If they notice it at all, it's usually beneath the level of political awareness. That lack of a public persona is a problem for the Bear because, like other rivers in the West, it's threatened. The Bear is under siege as more and more people ask their rivers to do more and more, even when its source and energy are finite.

Yet the Bear is unique, too, because it meanders through a particular landscape. That landscape has a peculiar geology, hydrology, and ecology. The history of human stakeholders of the Bear River is different, and that history provides us with a narrative of how we've used and misused the river. Reading that story critically gives us the opportunity to heal our relationship with the river and the environment it serves.

So the Bear River has several parts to play in its role as protagonist. At one and the same time, it speaks for all rivers, yet also just one. We can understand those riverine voices because they sing in harmony. The voice of all rivers reverberates in the Bear, and it wisely suggests that we listen.

Never dormant, always alive, the Bear River begins its five-hundred-mile long journey in the Uinta Mountains, emptying into Great Salt Lake.

Skies weep
My headwaters catch the rain,

Shy rills
Glisten on rocks

Or slip below the surface
And fall unseen.

I sculpt the land,
Wed earth in my bed,
Meandering to an eternal chord.

Abrading and braiding
I never reach the ocean.

I rest instead
In an inland sea.

At times, the Bear's cascades become waterfalls, its tributaries plummeting over shelves at right angles, as on Ostler Fork.

At about 11,000 feet, McPheters Lake is one of the highest in the Stillwater Fork drainage of the Bear River Basin.

2

The Ancient Saga of Water and Land: Geomorphology and Hydrology of the Bear

Like all rivers, the Bear begins in the sky. A molecule of water hovers in a gathering storm, anxiously suspended until the cloud no longer can support its weight. It falls and begins the earthly leg of its hydrologic journey.

If the land where that raindrop fell were a flat plain, its course in that river would be complex and full of promise. The hydrologist Luna Leopold showed that if you take random walks in any direction in a 180-degree arc, all paths eventually come together. River channels can form that same way, haphazardly on level ground. But when rain falls in high country where slopes are steep, water's path is more resolute. When that drop of moisture strikes 12,720-foot Lamotte Peak in Utah's Uinta Mountains, the mountain claims it for its river. There you'll find the highest headwaters of the Bear.

Still, those headwaters can't be called a "river." Most watercourses never reach that state of maturity. To become the headwaters of a river, that itinerant drop of water must not have its path thwarted by a pebble and sink into the earth instead. Only when it merges with other water droplets on the surface does it become the lowest order of a watercourse: a rill. In the Uintas, that is an adventure because the gradients are steep, and a newborn rill must plunge recklessly down the crumbling slopes of one of the oldest ranges in the Rocky Mountains, following the irrepressible force of gravity. As the trickle gains volume, the rill grows into a creek. Then creeks become streams as the earth channels more inflow. Finally, when these tributaries merge about ten different times, the watercourse fills out its body, and hydrologists using the Horton Classification System give it its final name: river.

Unlike a drop of rain falling on a midwestern plain where there is only one way out—to the Gulf of Mexico—when moisture lands in the steep, western Uintas, mere inches determine its final resting place. If you look at a map and trace the line of peaks and ridges, first in a southerly direction from A-1 to Kletting to Hayden Peaks and then Mt. Agassiz, then east/northeast to Spread Eagle and Yard Peaks, and finally north to Mt. Beulah, you've defined the southernmost boundary of the Bear River watershed. If a drop of moisture falls on the eastern or

As the Bear River deposited its load of fine sands and silt into Lake Bonneville, the deposition began to creep upstream, becoming the fertile farmlands of present Gem Valley, Idaho.

southeastern side of that ridgeline, it will wind up in the Green/Colorado river system and eventually find its way to the sea. If it falls north of that ridgeline, it becomes part of the Bear River Basin and earns a different fate. Instead of reaching the ocean, it becomes part of a landlocked river system. With a length of five hundred miles, the Bear is one of the longest rivers never to reach the sea. Instead, it empties into a great inland sump, where it is the primary nurturer of one of the world's richest, most unique ecological systems—the Great Salt Lake.

The Bear upsets preconceptions about a river in still another way. Most of us think of a river as a steadily thickening, linear thread stretching across the land from one point to another far away. But while the Bear River is more than five hundred miles long—beginning in Utah and passing through Wyoming, back into Utah, then into Idaho, and finally back to Utah again—its headwaters are only about seventy-five miles from its mouth. In fact, the Bear River begins and ends at roughly the same 40° 41′ 45″ latitude. Striking north, it makes a great U-turn 250 miles out near Soda Springs, Idaho, and then heads back as if it doesn't want to stray too far from home.

About 34,000 years ago lava extrusions created a great wall that prevented the Bear River from flowing north into the Snake River and then into the Pacific Ocean. It turned south at Soda Point and began emptying into Lake Bonneville.

Geologic Prehistory

A river is more than simply running water. It is both the route that water takes as it moves downhill as well as the medium that carries the bits of the land it erodes along the way. Dissolved or suspended in every river is the geologic history of the land in its drain-

The exposed benches of Lake Bonneville either become prime real estate, like the mouth of Logan Canyon, or they erode, moving the silts that once were the bed of Lake Bonneville back into the Bear River, their original source.

age area. A river and the land it passes through, then, are intimately entwined. In the case of the Bear River and its watershed, that relationship is complex.

In the southernmost reaches of the river, its headwaters in the Uinta Mountains, the rocks are very old: Precambrian dating from 570 million years ago and earlier. Farther north in the watershed is the Bear River Range, which bisects the Bear River Basin. A section of the Wasatch Mountains, the range is in the Middle Rocky Mountain physiographic province. According to geologist Don Mabey, these mountains uplifted about 50 million years ago when the underlying Overthrust Belt, a seismically active geologic province characterized by faulting and folding, caused older Paleozoic and Mesozoic rocks to ride up and over the younger Tertiary and Quaternary sediments.

In the recent Pleistocene, the Bear River's headwaters were glacial ice, and glacial scouring is evident in the catchment basins at 10,000 to 11,000 feet in the High Uintas. Great, smooth bowls carved from the rock act like funnels to collect water and send it down the Bear River watercourse. In his study of the hydrogeochemistry of the Bear's headwaters region, Michael Leschin determined those Quaternary-period glaciations stripped the basins to the Precambrian bedrock, dense rocks with low porosity and permeability.

The Wellsville and Clarkston mountain ranges west of Logan are largely marine sedimentary rocks from Paleozoic periods laid down from 525 to 280 million years ago. Trilobites left their mementoes discretely hidden in rocks of the Wellsvilles, according to geologists J.P. Evans and R.Q. Oaks.

At the northernmost reach of the Bear, before it turns south at Soda Springs, there was an inland sea 250 million years ago. It left its mark with the world's largest concentration of phosphate deposits. An arm of the ancient sea submerged a shallow shelf around what is now the town of Soda Springs, Idaho. Mabey notes that an upwelling of warm waters rich in organic life led to a two-hundred-foot-deep deposition of phosphate-dense shale.

When the Bear does make its turn at Soda Springs, it enters the Great Basin section of the Basin and Range physiographic province. Also an unstable area, this province of north-south fault-block mountain ranges is characterized by deep valleys of sediments from the relatively recent Quaternary period—1.8 to 1 million years ago; the alluvium eroded from adjacent mountain ranges like the western slope of the Bear River Range. These valleys can tilt from side to side like bongo boards, causing rivers to change course.

North of the Bear River is the Columbia-Snake River Plateau physiographic province, a broad lava plain recently formed. It's just beyond the reach of the Bear, but its geologic history is also tied to the river because the Bear at one time flowed north into the Columbia system and reached the sea. But recent cataclysmic vulcanism turned the Bear from a tributary to a river in its own right. It was cut off from the sea and its resting spot, but it gained a

During the wet, pluvial period of the last Ice Age, Lake Bonneville could take no more inflow from the swollen Bear River. It broke through an ice dam near Downey, Idaho, about 14,500 years ago and sent a cataclysmic rush of water north, carving Red Rock Pass.

unique identity in return.

The bottom line: it's difficult, if not impossible, to say how old the Bear River is. If you date it by the land it carries, it is as old as some of the oldest rocks on the continent. If you date it by its watercourse, it's young. If you date it from the creation of water and the hydrologic cycle during Earth's first volcanic period in the Azoic era more than 4 billion years ago, the Bear is older than its dirt.

Geomorphology

Besides carrying Earth's geologic history, rivers sculpt the surface of the land as they flow inexorably to lake or sea. The most evident sign of the Bear River in the recent geologic past is its interaction with Lake Bonneville. That geomorphic interplay also points to another unique quality of the Bear River: it is a river of two ages. While the upper half of its channel is middle aged for a river, the lower half is just a newborn in geologic time.

About 5 million years ago as the mountains around Soda Springs rose up, they simultaneously began slowly to erode. At one time, the gravel, sand, and silt from these ranges eroded into Lake Bonneville, an inland sea like other pluvial-era lakes that had formed during a period of increased precipitation. Eventually these eroded materials formed a divide between the subsiding Snake River Plain and Lake Bonneville.

About 34,000 years ago, extrusions of lava from vents around Soda Springs blocked the Bear River from flowing north along its course through Portneuf Canyon into the Snake River. The Bear was forced to veer south and began to flow into Lake Bonneville instead, emptying into a bay that is now Cache Valley in Idaho and Utah.

The effect on Lake Bonneville was cataclysmic because the Bear was a much larger river then. The geologist Grove Karl Gilbert estimated that one-half of the inflow into Lake Bonneville came from the Bear River drainage. On a daily basis, the Bear deposited tons of sediment into its estuary in the still waters of the lake. Soon the river filled completely with sediment upstream to a point around Grace, Idaho. These sediment deposits became the fertile farmland of lower Gem Valley.

Eventually Lake Bonneville could take no more inflow from the Bear. Approximately 14,500 years ago, Lake Bonneville broke through an ice dam near Downey, Idaho, and its waters raged north through Marsh Valley into the Snake River Plain. The wall of water cut what is now called Red Rock Pass.

Most of the river valley downstream from Bear Lake consists of fine silts and clays because of the earlier geomorphic interplay between the Bear River and the lake. Typically, because they are heavier, the coarse sands and gravels a river carries are deposited in the mouth of its delta, and fine sands and silts are carried farther into the lake or ocean. But the Bear River at one time in the recent past flowed into Bear Lake, a much larger lake then, more than forty miles long. These coarse sands and gravels were left upstream in Bear Lake. So only the finer sediments were carried downstream.

This deposition created a broad, braided delta where the Bear River entered Lake Bonneville, eventually covering Cache Valley. These became the Provo deposits of the Lake Bonneville shoreline. More than two hundred feet thick, they consist of the fine sands, silts, and clays that make the benchlands of Cache Valley so fertile—and potentially so vulnerable to erosion, as Calvin Clyde noted in his 1953 thesis.

Eventually, as the climate warmed and evaporation began to shrink Lake Bonneville, the Bear River started to cut into the exposed Provo deposits. Now the Provo level of Lake Bonneville is a terrace or bench some 180 feet above the current river plain,which averages about 1,600 feet wide in its lower reaches.

Because this stretch of the river through Gem Valley in Idaho to Great Salt Lake is relatively young, the riverbanks tend to be steep, much steeper than the usual angle of repose for mature rivers. Coupled with the fine sediments left over from deposition in the bays of Lake Bonneville, the banks are vulnerable to sloughing off into the river when they become wet. Many of these slide areas are active from Preston, Idaho, upstream to Riverdale.

Ryder Lake, one of the headwaters lakes in the Bear River Basin that collect water trickling into glacial bowls.

Small waterfalls, three to five feet high, cascade over a series of shelves above Ryder Lake, stacked like risers in a giant's staircase.

Geography of the Bear River Basin

Today the Bear River drains a basin of approximately 4.8 million acres or 7,320 square miles. It drops approximately 8,500 feet from its headwaters to its mouth at 4,200 feet in Bear River Bay in Great Salt Lake. There are 155 lakes and reservoirs in the basin, the latter impoundments symbolizing the newest interruption in the flow of the Bear, a political intervention that in some ways is as powerful in changing the course and soul of the river as earlier movements of the Earth.

The Bear River traverses more than just land. It crosses a plethora of governmental and private jurisdictions, and each claims a share of the river's lifeblood, sometimes competing with other political subdivisions, sometimes with the silent stakeholders of the Bear—the flora and fauna.

Consider the labyrinth through which the Bear River must flow. It passes through nine counties in three states. It begins in the jurisdiction of the Wasatch National Forest in the Uinta Mountains. From there it flows largely through private land, but along the way, it abuts state lands in Utah, Wyoming, and Idaho. The Bureau of Land Management (BLM) administers many of the acres adjacent to the river in Wyoming and Idaho. Finally, the U.S. Fish and Wildlife Service (FWS) controls the area at the river's mouth in Bear River Migratory Bird Refuge, as well as being a player with Bear Lake National Wildlife Refuge in the complicated water politics of Bear Lake.

The Bear River has approximately fifty tributaries, some needing to please no more than one governmental jurisdiction while others flow through the competing politics of national, state, or local interests. Then there are nongovernmental agencies like the Bear River Watershed Council and the Bear River Resource Conservation and Development Corporation that were set up to help political and ownership jurisdictions talk to each other in more constructive ways.

Different governmental entities define the Bear River Basin in a variety of ways. Those definitions, in turn, reflect the biases of various, humancentric perspectives. For instance, the Bear River Compact, the three-state agreement that partitions water rights, divides the basin into three regions: the Upper Bear, which encompasses lands from the Forest Service boundary to Cokeville, Wyoming, and uses that are primarily agricultural and grazing; the Central Bear from Pixley Dam to Stewart Dam, which includes Bear Lake and the beginnings of hydroelectric power generation; and the Lower Bear from Stewart Dam to Great Salt Lake, the longest reach of the river and the place where most new water development is poised to take place. One of the reasons for dividing the Bear into these three regions was to allow water upstream from Bear Lake to be developed without threat of litigation.

The Utah Division of Water Quality has a slightly different take. It divides the Bear River Basin into three segments based on "major

Water still has remarkable clarity in the high-elevation tributaries that form the Bear River. Underwater rocks are almost as sharp and distinct as those that breach the surface.

valleys." Even the Environmental Protection Agency (EPA) looks at the river in a slightly different way, preferring to see the basin as five different watersheds, a more logical approach that honors the river as a collection of drainages and tributaries, rather than entries in different political documents. The following description of the Bear River Basin adopts a geographic perspective.

Mountain Reaches: Utah

While the Bear River falls 8,500 feet from the top of Lamotte Peak to the Bear River Bay, more than half that drop occurs in just the first twenty miles. That steeper gradient, in conjunction with more typical headwaters geology than the sedimentary relics of Lake Bonneville, shapes the character of the river. There are five major tributaries of the Bear in the Uinta Mountains: East Fork, Stillwater Fork, Main Fork, Hayden Fork, and West Fork, a relatively small stream that's often included largely because of its storage potential. In turn, two of those tributaries have significant forks: East Fork is formed from the merger of Left Hand Fork and Right Hand Fork, odd names for streams when cardinal orientations have already been used; and Ostler Fork is a major tributary of the Stillwater.

Ryder Lake, the source of Stillwater Fork, is typical of the high mountain lakes that are part of the Bear River system. It fills a twenty-five-acre depression at 10,600 feet at the base of Hayden Peak, a well-worn mountain that looks more like a ridgeline and lacks the jagged features of newer ranges in the Rocky Mountains to the east in Colorado.

You hear the Bear River trickle into Ryder Lake before you can see it. Just below the last patch of July snow in the crumbling, unstable glacial cirque of Hayden Peak, drops from the previous winter's moisture and violent summer thunderstorms collect in hidden, tiny rills percolating between the cracks in the talus slope. Once the water comes to the surface, it moves over flat boulders, becoming a three-inch-deep clear, glistening creek as it slides effortlessly over the smooth rocks. The only sound it makes occurs when it gently drops from one tier to the next. Occasionally, when the boulders separate enough to allow granulated soil to collect, the first mountain meadows appear.

Most lakes in the Bear River Basin have unceremonious monikers. They are christened with the initials of the river, followed by a number, like BR 16.

On the east side of Ryder Lake is a gentler slope marked by a thick stand of conifers. To the west is a series of shelves, ranging in width from 50 to 150 yards, each shelf approximately twenty-five to fifty feet high and scalable. The shelves consist of thin soils resting on hard, glacial moraine and are covered with abundant meadows and ponds, the latter usually brimful even in drought years.

Actually Ryder Lake isn't the only lake in Middle Basin on the Stillwater drainage, nor is it the highest one. Unnamed lakes abound in the Bear River Basin, identified only by letters and a number, like BR 16, if they have any designation at all. Adding insult to injury,

The high, rocky, alpine meadows above Allsop Lake act as a collector system, moving snowmelt and water from seeps five hundred feet down to the lake inlet.

The late afternoon sun makes its mark, christening the outlet of Ryder Lake.

The Allsop Lake outlet marks the beginning of Left Hand Fork of the Bear River.

lakes often are christened with the names of humans, instead of after river features.

Four hundred feet above Ryder Lake sits McPheters Lake. Marked by boulders and polished, hard plates, the alpine landscape at 11,000 feet is sparse and obdurate. Meadows are small, barely clinging to limestone and dolomite, and great boulders several shoulders high loom on the scoured rock like thrones.

A series of stunning cascades and five-foot waterfalls trickle down the shelves from ponds at the same altitude as McPheters Lake and empty into the west side of Ryder Lake. The waterfalls are pristine. In fact, the clarity of water in the Bear River is remarkable at all elevations in the Uinta Mountains. You can look directly through the water to small cobbles, gravel, and sand as if there is no veil of secrecy, no need to disguise a lifeless bottom as in the lower reaches of the Bear.

The Allsop Lake outlet is fed by streamside springs, as well as the lake itself.

In the East Fork drainage, Allsop, Priord, and Norice lakes fill the glacial depressions in north-south orientations. Like Ryder Lake, as much or more water seems to come into Allsop Lake via a system of springs as from the inlet itself. At Allsop Lake, springs flow into the lake and its outlet via fifty- to one-hundred-foot-long rivulets every thirty to fifty yards, and an underwater stream extends from the inlet approximately three hundred yards into the lake until lake depth makes it difficult to see.

Hell Hole Lake and its outlet, the source of the main branch of the Bear River.

When water first leaves Ryder and Allsop lakes and begins its descent down the Stillwater and East Fork tributaries, it seems to take its time, content to meander gently between boulders that have become more spread out. But at 9,400 feet, Stillwater Fork changes character dramatically as it begins to fall precipitously. Intermittent cascades and waterfalls accelerate the stream velocity, and cobbles scrape against equally hard sheets of rock, making them even smoother before they collect in pools below the falls.

Once Stillwater Fork ceases cascading and reaches its lower stretches in the Uinta Mountains, it turns placid again and begins to meander easily through broad meadows. Eventually it flows through the popular Christmas Meadows area, where humans begin to claim ownership and dominate use.

Upper Flatland Reaches: Wyoming, Utah, and Idaho

Rivers worldwide behave remarkably the same because they seek a steady, most-probable state. Even when humans intervene in sometimes-misguided attempts to alter local conditions, rivers become contrary and tend to move back to their natural state. In flatland reaches, especially, that means meandering.

As Luna Leopold explained in his classic *A View of the River*, a river must convert potential energy at its upstream end to kinetic energy as gravity pulls its waters downstream. That generates heat, and that heat carries sediment and erodes riverbanks and beds. Since it has to follow physical laws as it utilizes that energy, a river also must conserve heat and mass. Because a river is an open system, it tends toward two conditions: a minimum of work and a uniform distribution of that work. Both conditions can't be sustained at the same time, so a river seeks a state of Zen-like balance, naturally moving to a state that eases the conflict. That compromise takes a physical form: a river's rhythmic meandering.

At forty-two acres, Amethyst Lake is the largest in the Bear River Basin.

Most stretches of rivers are sinuous or meandering. In fact, there is something about a straight line that aggravates rivers. They avoid it whenever possible. Moreover, rivers and streams worldwide uncannily mimic each other in their meandering to the point that, from a bird's-eye view and without a physical feature on the ground to provide scale, you cannot tell the difference between a small stream curving through its floodplain and a great river. They look the same.

The distance from one meander to the next consists of two opposing curves and is called a *wavelength*. A river also has a channel width that remains relatively constant over short stretches, although it slowly increases as more water enters the river as it moves downstream. Together, these figures measure the magic of meandering.

Straight stretches of a river are seldom more than 20 times the channel width. A wavelength averages about 11 times the channel width, and the radius of river bends is about 2.3 times the channel width. A river will simply have it no other way. It is the most

In the wet meadows surrounding Hell Hole Lake, the Bear begins to move, gently meandering, forming sinuous curves.

High mountain tributaries of the Bear can't meander for long. As the grade quickly becomes steeper, streams like Left Hand Fork start to fall.

common state of a river, and no amount of contrivance can permanently convince a river otherwise. In fact, this same sine-generated curvature in river meandering is duplicated in surface rivulets formed by melting glacial ice and even underwater currents like the Gulf Stream.

The Bear River begins its graceful meandering when it reaches Christmas Meadows in the Uinta Mountains. That pattern is even more prominent once the Bear leaves the Wasatch National Forest and enters the jurisdictions of Utah and Wyoming. That meandering also creates a wider floodplain, and when the river spills over its banks, on average every year and a half, it deposits rich soils.

When the Bear reaches the flats of Christmas Meadows at 8,800 feet, it catches its breath and meanders again.

Once the Bear River leaves the Wasatch National Forest, it begins to cut a broad floodplain through Wyoming, and the sky becomes a noticeable complement to the flatter landscape.

The rich floodplain of the Bear upstream from Evanston, Wyoming, has become the province of private property. Barbed wire tries to sever the river.

That is bottomland, and human beings covet it. At that point, lands adjacent to the Bear River are in private hands, and in Utah, ownership of land includes the riverbed.

The Bear follows this gently meandering course through Utah, Wyoming, and the Bear Lake Valley in Idaho with little deviation. Only a diversion near Bear Lake interrupts that riverine harmony. Two major tributaries enter the river—Smith's Fork in Wyoming and Thomas Fork in Idaho—but for the most part, the Bear is on its own.

When the river changes its course at Soda Springs and heads back south, the geography changes markedly. Instead of a broad floodplain, the river begins cutting a canyon approximately 350 feet deep below the dam at Soda Point, the northernmost extent of the Bear River Range. The river has little room to move because it becomes a young river again, and like a teenager wanting to break out, it cuts into the base of the mountain range or tries to undermine the sheer basalt cliffs on its north and west banks left when vulcanism took a hiatus and deposited its cooled-magma calling card 34,000 years ago.

Those steep basalt cliffs christen this stretch of the Bear the Black Canyon from Soda Point downstream to a point just below Grace, Idaho. Looking down on the Bear from the cliffs near the Turner-Grace Bridge reveals a striking change of tonalities in its geography. The canyon walls are a deep black, intermittently laced with

Small, clear-water tributaries like Stauffer-Nounan Creek in Bear Lake Valley provide feeding and spawning habitat when private landowners gracefully acquiesce to the beaver's civil engineering.

streaks of bleached white rock. At the bottom of the canyon, the river contains only a streamlike amount of water—looking as if it is in a state of permanent drought—only because most of the water has been taken out upstream to generate hydropower. The river bottom is bleached white travertine, a rock formed from calcium carbonate that leaches from deposits of limestone upstream or enters the river via springs. A ribbon of clear blue water in Black Canyon curves and dips from one sculpted pothole to another as if it's being ladled.

Lower Flatland Reaches: Idaho and Utah

After the Bear leaves the narrow confines of Black Canyon at Grace, it again begins its leisurely meandering through the southern end of Gem Valley. Another dam momentarily backs up its waters at Oneida Narrows before it continues its course through northern Cache Valley in Idaho and into Utah. Occasionally, in these broad valleys, there are oxbows that formed when the river changed course due to high water or drought and a bend was cut off from the main channel. These biologically fertile remnants of the river become marshes when they are continually replenished with subsurface water.

In long reaches of the Bear through Black Canyon, the river is reduced to a crystal-clear, spring-fed trickle because most of its water has been diverted upstream for hydropower generation.

In Cache Valley, four main tributaries—the Cub River, Logan River, Blacksmith Fork, and the Little Bear River—augment the main stream by about 50 percent. Then the river backs up behind another hydroelectric dam at Cutler Narrows. Below the dam, the Malad River enters the Bear in Box Elder County before it begins its last run through the farm and ranch lands of northern Utah and

Boulders breaking off the steep walls of Black Canyon create rapids in the Bear River.

As the Bear meanders through Gem Valley, Idaho, leveling a broad swath, it creates incipient oxbows like the one on the left. Eventually these oxbows are cut off from the river by sediment deposition or a change in its course due to flooding or drought.

empties into Great Salt Lake. But to say that the river is *running* at that point is stretching the metaphor.

At its mouth in the Bear River Migratory Bird Refuge, the river assumes its last geomorphic stage: braiding. A river braids because the sediment load exceeds its carrying capacity and those sediments collect on the riverbed. Eventually the deposits block a river's flow and split it into smaller branches. When a river braids, it is a tired river, and the Bear, as it passes through its eponymous refuge, is more tired than most. Its waters have been used and reused countless times to irrigate and generate power. Foam on the river's surface has apparent motion only if you look away and glance back sometime later. Then a glob of foam will appear to have moved one iota, and the river says, "I'm still alive and moving," but you also know that its end is near as it makes its last contribution—replenishing Great Salt Lake.

Hydrology

More water falls on land from precipitation than is absorbed back into the sky from evaporation or transpiration from plants, approximately 33 percent more. That excess water is what flows in rivers, whether surface or subsurface.

Precipitation in the Bear River Basin varies dramatically. The headwaters can receive more than 40 inches of moisture in an average year, mostly in the form of snow. But at its mouth in Great Salt Lake, the mean precipitation is only 9 inches. The average annual precipitation for the basin is 13.6 inches. Wyoming rangelands receive 12.5 inches, but during the growing season of May to September, arable lands get approximately 5 to 6 inches of rainfall when they need 20 to 30 inches for growth.

The average annual temperature of the Bear River Basin is forty-three degrees , but that doesn't tell us much since the extremes are so great. Summer temperatures can top a hundred degrees, and radio-reporting weather stations in some sinks in the Bear River Range have recorded temperatures sixty below and colder. One thing that remains constant is that most moisture in the basin is seasonal, coming in winter, and runoff reflects that, although monthly flows can vary. The first runoff in early spring comes from snowmelt in the valley bottoms, while the second peak in June comes from melting snows in higher elevations.

The hydrology of the Bear becomes even more complex, with the potential for confounding water claims, when the basin is broken down into its ten hydrologic subbasins. The largest is the Cache subbasin, and it is almost twice the size and contributes more than twice the runoff of any of the others.

It is important to note, too, that the Bear, like all rivers, has a sibling that runs underground. All basins have rills, creeks, streams, and rivers on the surface, even if they only run intermittently during the year. But not all water that falls from the sky and is collected in

Portions of the riverbed in Black Canyon aren't black at all. Instead, the river has sculpted great pools out of travertine, leaving what looks like a string of exotic bathtubs.

headwaters winds up in surface flow. Much of that moisture—from 5 to 50 percent, depending upon climate and land use—percolates down into the rock, where it saturates the tiny spaces between alluvial material like sand, gravel, silt, and clay, or the crevices of fractured rocks, and becomes groundwater.

There is direct interaction, too, between surface and subsurface water. In fact, about 40 percent of water flow in rivers nationwide comes from groundwater. This groundwater can enter a river laterally below its surface, or from above through a seep or spring. Similarly a river can give up some of its surface flow to the surrounding groundwater as it infiltrates into the adjacent alluvium.

Most groundwater is located in aquifers, underground layers of porous rock where water collects from hydraulic gradients that run downhill like rivers. These aquifers can be confined or unconfined; the former are sealed off by layers of impermeable clay and consequently under pressure, while the latter are more common. Unconfined aquifers are permeable, and water moves into and out of them with ease. They make up the water table, which is close enough to the surface of the land to be affected by atmospheric pressure and

At the southern end of Gem Valley, the Bear River becomes a reservoir as it begins to back up behind Oneida Narrows Dam.

As the Bear leaves Oneida Narrows, it runs clear again, having deposited its silts in Oneida Reservoir.

(above) In winter, the Bear pauses long enough to transform itself into intricate geometries of ice. Still, paw prints reveal that life moves along the frozen river.

(below) Most precipitation in the Bear River Basin falls as snow. In the Uinta Mountains, the Bear disappears under a blanket of snow and ice.

(above) Fog is common in areas adjacent to Cutler Reservoir in Cache Valley in late fall.

(below) When the Bear becomes a tailwater river below reservoirs like Oneida Narrows, it moves free of ice.

changing hydrologic conditions. Groundwater is a major source of drinking and irrigation water in the Bear River Basin.

In the headwaters region of the Bear River, the rocks are relatively nonporous, so groundwater tends to flow in the fractures between the Precambrian rocks. Like the river itself, groundwater in the mountain region tends to flow north. Groundwater also follows the same divides in the basin that cause the surface water to flow into one tributary or another.

Just as the steep slopes of the headwaters cirques are important to the surface flow of the streams that form the Bear in the Uintas, Michael Leschin noted that the hydraulic gradient of groundwater in the mountains affects the stream chemistry. The steep gradient in the cirques causes the headwaters of the Bear River to be relatively dilute because the water has less time in contact with the fairly insoluble minerals of the quartzite terrane. Stillwater Fork has the steepest hydraulic gradient of all the upper tributaries but only by a few percentage points.

Moreover, because the subsurface flow runs parallel to the river between impermeable rocks, dissolved minerals from carbonate rocks like limestone don't enter until far downstream. Some upwelling of dilute water, too, via capillary action, further isolates the upper river from dissolved carbonates that wind up congregating in the Black Canyon as travertine.

Groundwater works its way to the surface near Cokeville, Wyoming, in deep blue potholes.

According to a 1997 study by the Utah Department of Natural Resources, snowfall in the Uinta Mountains contains few dissolved ions. So most dissolved ions in the river come from exposure to surface earth materials. That means that the minerals in the headwaters are mostly quartz, feldspar (which is the primary component of the clay sediments in the river downstream), some calcite from limestone, and dolomite, plus traces of hematite, muscovite, and illite, a claylike substance.

The Bear River is mostly basic and rarely acidic. The headwaters have a pH of 7.3 to 7.5, with Stillwater Fork dipping to mildly acidic 6.7. As the waters move downstream, they become more alkaline, and the figures change considerably.

Salinity is about 100 milligrams per liter in the headwaters. Where the river reenters Utah from Idaho, the figure jumps to 560 milligrams per liter. At the confluence with the severely saline Malad, its last major tributary, salinity in the Bear increases to 800 to 900 milligrams per liter.

Periodically the groundwater along the Bear shows itself in streamside springs. Some of these springs flow directly into the river. Others are contained as potholes.

Northern Bear Lake Valley has significant geothermal activity. Near Soda Point, hot subsurface water flows through limestone and dolomite before it mixes with colder groundwater. That dissolved

There is significant geothermal activity adjacent to the river on the west side of the Bear River Range. Hot water and steam hiss and burp both in the river and along its banks just upstream from Oneida Narrows Reservoir.

The interaction of steep banks, formed by silts left from Lake Bonneville, and groundwater hydraulics causes embankments to slough into the Bear, creating reaches of murky brown water.

limestone and dolomite carry carbon dioxide gas, calcium, magnesium, and bicarbonate, a veritable mixture of patent medicine from the Old West. Where it burst to the surface under pressure, passing itinerants named the bubbling waters Soda Springs or Beer Springs. The town that developed along the emigrant trail took its name from the springs, but they no longer effervesce near Soda Springs, Idaho. Now they are buried under Alexander Reservoir.

But other hot springs do come to the surface periodically along stretches of the Bear, supplying elixir for tourist stops at Riverdale in Idaho and Crystal Springs in Utah. On Maple Grove Road just north of where the Bear flows into Oneida Reservoir, four hot springs hiss and burp continuously, both along the riverbank and from the streambed about thirty feet into the river, sending up plumes of steam. Calcium carbonate buildup has created colored rock formations like the ones in Mammoth Hot Springs.

Sedimentation

A river can carry both dissolved materials and insoluble particles, which are called *sediment.* Dissolved materials can vary from as little as 1 or 2 percent of the total load in arid areas up to 64 percent in wetter climes. The smallest particles of sediment are silts or clays, usually clay materials created by the chemical alteration of feldspars. Next in size are very fine sands, followed by medium sands, usually quartz crystals, which are hard to dissolve in water. Coarse sand consists of quartz crystals in the grains of other minerals. Particles more than a millimeter in size are bits of rock containing more than one mineral. These sediments fall into the category of larger rock fragments, starting with pebbles and increasing in size through gravel, cobbles (those smooth stones called *river rock),* and, finally, boulders.

Suspended load is fine material in transport that is thoroughly mixed with water. It's what makes a river look muddy, like the Bear River for much of its length. *Bedload,* on the other hand, is the larger particles like pebbles, gravel, and cobbles that are pushed along the riverbed. These sediments only move when the river is running higher than average, usually only a few days a year.

Because the Bear's headwaters are so clear, you can identify the primary sediment: bedload. Often these are the rounded stones near the outlets of lakes that kids inevitably fling laterally to see how many skips they can make. Such rocks in the headwaters of river systems like the Bear are larger due to jointing in rock outcrops nearby. They provide the large blocks that are the raw material for bedload. It is the product of past geologic activity, especially glaciation, and prevailing weathering conditions.

Downstream has more suspended load, largely due to abrasion. Rock fragments moving along the streambed or scraping against the sides of riverbanks break apart or become rounded. When a rock fragment becomes small enough, it is suspended in place;

Riffle-pool dynamics are evident in the stretch below Oneida Narrows. The exposed bedload is full of larger cobbles that were moved during periods of high water.

the velocity of the water is able to support its weight. A river's current sorts the fine particles, too, and coarser fragments fall to the streambed sooner. The finest particles can travel all the way to the river delta under average flow conditions.

The Bear River runs crystal clear in many areas besides its headwaters. There is a gravel bed from Oneida Reservoir almost as far downstream as Preston, Idaho, although intermittently the river becomes cloudy when a muddy tributary enters it, and since it takes time for suspended load to dissipate. But in the scant three miles just upstream from Preston, the river changes dramatically. Three tributaries enter along the west bench: Battle Creek, Deep Creek, and Dayton Creek. Tremendous amounts of sediment from these creeks flow into and cloud the Bear, so much material that it chokes the river into a narrow channel and commonly floods it onto adjacent fields, even during times of moderate discharge. Data suggest that much of the sedimentation from these streams is the result of increasing erosion upstream from land-use practices that have accelerated the movement of fine sediments left over from Lake Bonneville into the tributaries.

The effect on the Bear has been startling. As an illustration, while sediments deposited in the river below Oneida Narrows average about 68 tons per day, that figure jumps to 350 tons per day near Preston. Based on an earlier survey taking measurements from 1910 to 1950, that means that 20 million tons of sediment have entered the Bear. It has raised the streambed six feet, according to Calvin Clyde.

Sediment load can vary, too, depending upon the season. More sediment is carried during higher discharge periods in June and July. On two different days during the water year from 1990 to 1991, sediment load in the Bear River below Cutler Narrows varied from 103,000 kilograms a day to 325,000 kilograms a day. In some stretches of the Bear, especially when it's moving slowly, it looks as if you can walk across the river.

Erosion

Erosion in the Bear River Basin is more from physical than chemical weathering since there is little carbonic acid in the river, the primary agent of chemical weathering. But the river is its own agent of erosion. In fact, as in all rivers, there is an equilibrium between the erosion from one concave bank, where the river wears away bank material, to the downstream convex bank, where the river deposits the recently abraded material, because the velocity of a river slows at the inside of a curve, and the current can't carry the sediment, so it drops it in place. Luna Leopold summed it up in his classic phrase, "The river is the carpenter of its own edifice."

The Bear River has particularly severe erosion problems in some places due to its geomorphic history. Largely it's because of the lingering effects of Lake Bonneville and its different terraces as it rose

Rivers are dynamic hydrologic systems where velocities can change dramatically in short stretches. Each of those velocities, interacting with local soils, rocks, and obstructions, creates different microhabitats. (Key: The larger the arrowhead and stem, the faster the current.)

and fell. Most of the sediment left over from Lake Bonneville created layers of fine sands and silts, along with periodic layers of clay. When moisture from rain or irrigation percolates down through those layers, it can pass through the sands and silts but becomes blocked when it hits the clay. Then the water moves horizontally until it reaches the surface. There it becomes a seep. As more water follows that gradient, the sands tend to wash out, and the soils above become unstable; then the bank caves into the river.

The same problem occurs in the upland terraces, those higher benchlands left over from Lake Bonneville. Although these soils are fertile enough for agriculture, they also are fine sands and silts. Improper irrigation of those benchlands causes runoff that creates deep gullies, characterized by circular heads at their tops from material collapsing or sloughing off when clays prevent the irrigation water from percolating deeper into the aquifer. Then these tributaries, like the three just upstream from Preston—Battle Creek, Deep Creek, and Dayton Creek—carry and deposit tons of eroded material into the river, causing accelerating sedimentation of the Bear in several stretches. Without that accelerating erosion, the Bear would run clearer in some sections, and the bed would be more exposed pebbles and cobbles since finer particulates tend to gravitate down through the larger rocks of the streambed.

The Bear is also susceptible to another kind of erosion. While it's accepted that a riverbank will erode during high runoff when there is more abrasion by suspended solids and water rubbing against the soil, that is not always the primary cause of erosion next to rivers. During high-water stages, water seeps into the riverbank. When the river's current begins to ebb, however, that moisture wants to flow back into the river along the same gradient. This creates pressure on the riverbank, and slumping due to wetting causes the bank to slough off into the river. This is especially a problem in semiarid areas, like the lower elevations of the Bear River, and where the riverbanks are relatively steep, like stretches of the Bear with fine sediment rather than gently sloping river rock.

The Bear's siltation has an effect on riffle-pool dynamics. Like the predictability of meandering, rivers provide another kind of rhythm: alternating between *riffles,* which are areas of shallow, fast-moving water with stones or gravel, and downstream pools of slower-moving water that have fine sediment bottoms. This eternal yin and yang of riffle and pool repeats at intervals of five to seven river widths in straight stretches. The same thing happens in meandering stretches, but these intervals are tied to the phenomenon of alternating concave and convex curves.

This alternating riffle and pool structure is less pronounced in the flatland stretches of the Bear River, where its bed is composed of silts and fine sand. Especially in channeled sections, the Bear's riverbed has become uniformly silty. Coupled with increased erosion from runoff over ancient Lake Bonneville deposits, the silt wreaks havoc on some aquatic insects, which need riffles for their larval stage—leading to a loss of biodiversity in the river.

Velocity

Each river has its own personality, at least in the way that it responds to local conditions. Peak discharges happen at different times of the year. Riverbeds vary in color, based on the types of rocks in their basins. Depending upon local climate and plant life, riverside flora costumes a river with different dress, changing with the seasons.

The surface of any particular river looks different, too, and reveals distinct irregularities in the way that moving water interacts with the immediate environment. Eddies form around boulders that come to momentary rest in the middle of a river. Currents reverse themselves when a river cuts deeply into a bank, creating a kind of curved scoop that propels water flow backward. Underwater obstructions like logs or discharging springs create upwellings. Water taken out of a river for irrigation can also find its way back into the stream, changing the local conditions, sometimes in harmful ways.

But all rivers must bend to the laws of physics, and that makes them resolute in their common, almost-obsessive drive toward a predictable, steady state. That is also true of river velocities. There is little variance: all rivers tend to flow faster as their waters move downstream. Velocities are the same for all small rivers, running from a half foot to five feet per second. Large rivers run faster, from two to seven feet per second.

Any fisher knows that velocities change. Knowledge of those speed variations helps the fisher know what insects may or may not be present and how to present the fly or lure in a way that seems to

The Bear has multiple personalities. In its steeper mountain regions, it is exuberant and raucous...

flow naturally with the current. For instance, river velocities slow at the banks due to friction with the land. Similarly the current is slower along the riverbed because the gravels and cobbles provide friction. Also each layer of water acts as a drag on the level below it. All told, that means rivers run faster at their centers and on the surface.

Oddly, in their straight stretches, rivers are also somewhat higher in their centerlines during peak flow. That's because they have two circular cells, one at each bank. Water flows from the middle of the river to the bank, where it dips down and runs along the riverbed and back to the centerline.

River velocities are so predictable that they serve as reliable engines for commerce and travel. During the Great Depression, homeless itinerants on the Mississippi anticipating the approaching winter constructed crude rafts on top of a single, large log. They had learned that rivers run faster in the middle, so the log under the raft inevitably found its way to the center of the river due to its faster velocity and the cross circulation of the flow. If the log/raft approached a curve, it bent with the meander and headed for the

. . . At other times it barely moves, pausing to reflect on its journey from land to sky and back.

next curve. As it moved from one curve to the next, it drifted back and forth on each side of the centerline, but it never collided with a bank as it followed the hypnotic, helical flow of the great river. While the Bear never has led to a land of milk and honey, during the late 1800s, timber companies relied on the hydrology of spring runoff and greater channel width to move logs down the river to waiting railroad cars.

The Bear River follows these inexorable laws, too. Because the topography changes, it falls faster in its upper stretches than it does in the flatland, where the slope tends to level out. In the Uintas, it rides like a roller coaster, crashing and thundering down its rock-hard channel with that kind of youthful exuberance that is common to mountain headwaters. There it is wild, scenic, and commanding.

But a river's geography takes its toll as the river continually butts its head against the land. Then there are the human interventions that tap the river's kinetic power or withdraw its life-giving elixir for thirsty crops. Along the way, the Bear supports untold numbers of creatures and hydrologic systems. The land and its dwellers would be impoverished without the river. So, by the time it reaches its resting place in Great Salt Lake, it deserves that rest. There, the river waits—until solar energy once again energizes it and draws it into the sky on its ancient path back to the headwaters.

Riparian habitats like this one in Black Canyon can be fertile and dynamic.

My tributaries trickle down
Through life
Beating
In niches
Where flora and fauna seed.

I water the graceful green and
gold — and brown
Transitions quaking in concert with
sun and plow.

Forested bluffs now blind
I spread through low desert,
My wetlands plenishing a song
Of refuge.

Columbines are the monarch wildflower in the basin. They are found in undamaged, wet environments from high to midaltitude elevations, although their need for specific pollination agents isolates them.

Rockfalls or human-engineered breaks create both slack-water and fast-water environments, which support different life stages of aquatic insects.

3

Chapter

From Alpine to Desert: The Changing Ecology of the Bear River Basin

As the Bear River Basin scribes its great, five-hundred-mile long looping arc, it passes through a stunning spread of vegetative zones. From alpine to desert the Bear is a major character in nature's ecological narrative. It remains a story with a biologically diverse cast, but authorship changes from the singular voice of nature to collaboration with human beings.

The habitats of the Bear River Basin are in flux. Recent geologic history is playing a slowly evolving role, especially in the aquatic habitats of the lower riverine system and its streamside communities. Human interventions are accelerating changes in the Bear River's ecology—sometimes passively, sometimes actively. Writing a quickening narrative, human beings are altering the environment in which all plant and animal stakeholders must try to survive.

Alpine, subalpine, mountain brush, upland, and desert shrub are the primary life zones characterizing the Bear River Basin, as well as most Intermountain West landscapes. But suggesting these vegetation zones are defined by crisply separated species would underplay the rich biological diversity and create a convenient but artificial portrait. Rather, transitions between each zone are subtle. Sometimes one zone thrives within another, dominant one. That's because the zones contain varied microenvironments created by intricate interactions of soils, hydrology, elevation, and climate. Something as simple as the land's angle to the sun can make a dramatic difference.

The upper reaches of the Bear River Basin are a classic montane region, marked by alpine, subalpine, and mountain-brush communities of flora and fauna that flourish in a higher, wetter landscape. Typically these life zones lie at elevations from 8,000 to 11,000 feet. But there are no crisp boundaries where a change in altitude slices off one habitat from another. In reality, the mixture of microhabitats is only secondarily defined by elevation. Other factors come into play.

For instance, while temperature is a function of elevation, mountains create some anomalies. Depressions like Peter Sink in the Bear River Range can trap cold air in the winter, leading to arcticlike climates with temperatures that can dip to sixty degrees below zero, creating colder habitats than those on higher slopes. Similarly cold air tends to tumble down steep mountain slopes and collect in lower eastern valleys, explaining why Rich County,

Tundralike organisms populate the alpine zone: lichens, grasses, forbs, sedges, and low-lying shrubs. Ranging widely through various elevations along riparian corridors in the basin, moose leave their calling cards.

Utah, and Caribou County, Idaho, through which the Bear flows, are often the coldest spots in the lower forty-eight states.

Mountains, so characteristic of the Bear River Basin, work with the sun to create wet and dry environments that are only marginally affected by elevation. North- and east-facing aspects of mountains tend to be wetter climates, holding the moisture in their soils longer. These slopes are better habitats for the conifers of the Bear River Basin: subalpine fir, Engelmann spruce, lodgepole pine, and Douglas fir. Depending upon how steep these slopes are and if the surface or groundwater flows are slowed enough, new microhabitats are created for a slightly different mix of montane flora and the insects and animals that rely on these plants.

South- and west-facing slopes in the Bear River Basin tend to lose their water faster to the hydrological cycle and have warmer, drier climates over more months than north- and east-facing aspects. That explains why sagebrush, a dominant plant in lower, drier elevations, can extend upward to almost 10,000 feet in the basin.

Soils, too, interact with temperature and elevation to create different ecological niches. The highest elevations of the Bear River Basin in the headwaters area, which are characterized by less

Elevations of 11,000 feet or higher in the Bear River Basin are the province of the alpine life zone.

oxygen, longer winters, biting winds, and poor, rocky, shallow soil, support only tundralike organisms: lichens, grasses, forbs, sedges, and low-lying shrubs. Lower in the basin, in the bottomlands or riparian areas close to the Bear River, soils change from the metamorphic and igneous rocks of the headwaters to weathered shale and sandstone. Moreover, the fine silts and clays left over from Lake Bonneville are relatively deep, retain water much longer, and tend to be moderately saline and strongly alkaline.

The montane habitats of the Bear River Basin also contain numerous high-mountain wetlands. These are the small lakes and ponds with no names other than the "BR" surname and a humble number attached. Often they are the discharge areas of clear mountain springs. Nevertheless, they are important environments for the plants and animals that rely on them. Field biologists like Daniel Vice and Terry Messmer cataloging these species report that marsh marigolds, water sedges, Canadian reed grass, and tufted hairgrass populate these wet environments. At lower elevations, wild iris and American vetch speckle the margins, creating abstract expressionist fields. Mule deer, elk, moose, and red fox rely on these areas to feed and rest. Cooper's hawks and goshawks glide on thermal uplifts looking for prey around the ponds. Yellow warblers and song sparrows hold operatic auditions. And it's always startling to find a pair of mallard ducks nesting or feeding on a lake some 6,000 feet above where you normally expect to see them.

These wetlands are attractive to cattle for the same reasons that wildlife enjoy them. Even as grazing allotments are being retired or willingly sold, some are still interspersed within the high mountain habitat. Because ranchers usually can't afford to manage (herd) livestock intensively on their distant, summer ranges, cattle tend to congregate around these wetland areas when they graze in the vicinity. Then the wetlands can be at risk due to overgrazing and subsequent erosion.

Aspen forests are probably more endearing than coniferous ones due to their changing character and pioneering attitudes. They tend to thrive in wetter areas of the Bear River Basin, largely in montane regions. But they also can be seen at lower elevations, especially along tributaries of the Bear, where soils pull moisture from the river. Aspen forests are more important to wildlife, too, than are coniferous ones. While conifers are a climax forest, they tend to foster a less diverse animal community because their dense shade prevents the growth of an understory of shrubs or grasses. Just as aspen forests provide a transition from montane environments to lower, wetter upland habitats, larger animals use their margins for safety zones and transportation corridors. Because aspen forests are open enough to allow the growth of rich, abundant grasses and shrubs, deer and elk move in and out of the trees to feed. When threatened, they can retreat into the cover of the forest.

When a coniferous forest is destroyed by wildfire, suckers from aspen groves move quickly into the nutrient-rich habitat and set

Shallow ponds provide rich wetland habitats even at 10,500 feet in the Bear River Basin.

up new forests to replace them. Animals follow them to exploit the new opportunities because the landscape becomes more open and dynamic. In this sense, aspens are probably the most active players on the montane stage of the Bear River Basin.

Plant communities become dormant in winter in the higher elevations of the Bear River headwaters. Black bears are hibernating, and the ground and red squirrels and chipmunks enter periods of low metabolic activity. Not so the moose. They come down from the wetlands of the high meadows to live along the riparian areas of the Bear River, feeding on the reddish shoots of Drummond's willow. Most congregate in a stretch along the Utah/Wyoming state line. Cougars and bobcats continue to prowl, their paw prints in the snow

Aspens are a pioneering tree and mark the transition between coniferous forests and upland habitats. After a wildfire, like this one on East Fork of the Bear River, aspen clones and fireweed are the first plants to repopulate the carbon-enriched landscape. The emerging aspen forest will foster a robust environment for more diverse species than coniferous forests do.

the only visible evidence people will likely see. Although threatened, the Canada lynx has been sighted in the headwaters area.

Mountain brush is part of the montane region of the Bear River Basin, populating lower mountain slopes or rolling uplands from 6,000 to 8,000 feet. The area is prime wildlife habitat as big game browses on bitterbrush, serviceberry, buckbrush, and chokecherry. Blooms from cliffrose dot the open areas in midsummer. Mountain mahogany wafts a pungent aroma, and Rocky Mountain maple slips on a warm-colored dress in the fall.

At slightly lower elevations in the Bear River Basin, upland habitat dominates. These hilly areas can be lower steppes as the mountains roll out toward the Bear River floodplain, or even bluffs overlooking

A mountain mahogany at Soda Point marks the transition between the mountain-brush and upland vegetation communities. Sagebrush can range from 5,000 to almost 10,000 feet elevation on south-facing slopes in the basin.

Beaver have made a comeback in the basin, and their environmental engineering provides habitat for a variety of other species.

the river. It's a drier environment of piñon pine and juniper trees. Where there are large enough openings among the piñons and junipers, big-toothed sagebrush thrives. And when it flourishes, so do the sharp-tailed grouse and the sage grouse, although the latter are increasingly threatened in areas of the inland West that are losing sagebrush to drought and human mismanagement.

Ecology and the Human Interface

Most of the montane region of the Bear River Basin falls under the public jurisdiction of the U.S. Forest Service. Except for some grazing allotments, the Forest Service does not have to manage many consumptive uses of the land in the Uintas and Wasatch National Forest. There is little mining in the headwaters, and timber cutting, especially in areas dominated by lodgepole pines, isn't profitable due to the limited number of board feet that can be carved from a tree with such a narrow circumference. So montane regions remain relatively pristine, and most of the human impact on the ecology results more from recreation than extractive industries.

Once the Bear River leaves the mountains, however, there are more significant ecological changes, partly due to different land-management practices. The BLM has jurisdiction over some lower-elevation lands. This agency can't be as protective of the natural ecology as the Forest Service is because the BLM is charged with facilitating grazing and mining. Ecological change is also partly due to the river's interface with private property. That interaction, political as much as physical, is leading to a loss of biodiversity. River ecology, then, is influenced by nonriverine forces.

Once the river reaches lower elevations at around 5,000 to 5,500 feet, the vegetation changes. Uplands start to become agricultural, and diversity gives way to the monoculture of farming. Farmers concentrate on raising hay and row crops, as well as cultivating some small grains. Many of these crops can grow only because river water is used for irrigation, which shrinks the biologically rich riparian zone next to the river. Then chemical and pesticide runoffs from the uplands and the flatter desert areas of the Bear's floodplain further alter river ecology.

The Bear River flowed directly into and out of Bear Lake until the Pleistocene, when the river separated from the lake. In 1911 engineers constructed an intake canal to convey river water back to the lake for storage and hydropower production, but that 11,000-year separation from the river allowed Bear Lake to evolve startling, blue-green, Caribbean-like waters. Because the lake is nutrient poor with little plankton, it is oligotrophic with a relatively high oxygen level. That makes the waters exceptionally clear. Although there isn't a lot of feed in the lake, these conditions have fostered the evolution of some unique creatures: the Bonneville cisco, a grunion-sized fish with similar habits, Bear Lake whitefish, and Bear Lake cutthroat trout. The lake also sustains a healthy

Ecological Challenge in Cache Valley

Alice Lindahl, professor of biology at Utah State University, has monitored the ecological health of Bear River riparian areas and floodplains in Cache Valley for decades. Her report card shows mixed grades and room for improvement.

"The biodiversity is altered from what it used to be. I'd say there's less than 10 percent of natural cottonwood/ willow riparian system still intact. Cutler Reservoir is nice habitat, but it isn't the riparian that used to be there. It's flat marshland vegetation," Lindahl reports.

It's not just a paucity of woody vegetation. There is a dearth of macroinvertebrates, too (insects, crustaceans, flatworms, mollusks). "They are really low," Lindahl continues. "I never take my classes collecting either on the Bear or on Cutler because the diversity is so slight, especially compared to irrigation ditches and Barrens' ponds. Almost every place is better than the Bear."

One reason there are meager populations of macroinvertebrates is fish. There still are a few warm-water fish present, as Arnold Bangerter discovered, and they are predaceous. But turbidity is the bigger problem. Insect larvae that have exposed, soft gills cannot survive when the sediment load is high; it interferes with their oxygen intake. High turbidity isn't good for ducks, either.

"The sediment prevents aquatic plans from growing on the bottom because it's too dark down there," Lindahl says. Areas around the river used to be good gadwall duck nesting habitat because there was so much for the parents to eat. But the increased siltation of the Bear in lowland areas in Cache County, due to upstream erosion from damaging farm practices, has ruined that nesting habitat. Carp haven't helped, either, because their feeding habits stir up the silty river bottom and prevent benthic, vascular plants from growing.

"Turbidity is worse for wildlife," Lindahl continues. "It's worse for diving ducks that need to eat fish. It's worse for dabbling ducks that need to get plants off the bottom. Most of the wildlife you see there are either migrating or resting, but they feed somewhere else. The species we have that are successful, like white-faced ibis, which are aquatic feeding birds, are actually feeding on the agricultural fields. That's where they get their nutrients.

population of snails and bivalve mollusks. But that is changing as more nutrients from human septic systems work their way into the lake, adding to the irrigation runoff from area farms that enters via the intake canal.

The Bear effectively ceases to flow at Stewart Dam, where most of the river is rerouted to the lake via the intake canal. Only a trickle seeps through headgates at the dam. The river flows again when water is released at the Bear Lake outlet for irrigation. Then five dams alter the ecology: Soda Point, Grace, Cove, Oneida, and Cutler. Spawning fish are interrupted in their migrations. Water-level fluctuations due to changes in diurnal and seasonal releases from dams for irrigation, rather than natural flow patterns, have dramatic effects on the riparian ecology. Streamside vegetation or insects must cope with alternately being inundated with water and

They don't get it from wetland areas. That's true for snowy egrets, too. They follow cows around and grab insects."

Exotic, nonindigenous species have changed the ecology of the Bear River riparian area, too. Sometimes the changes are a bit ironic. Take the case of Chinese ring-necked pheasants. They were introduced to provide upland game hunting. But their numbers are declining, partly due to predation from raccoons and red foxes, which also are species that are not indigenous to the Bear River Basin. It's exotics versus exotics, and ground-nesting birds tend to lose.

Agricultural feedlots along the Bear are encouraging the proliferation of starlings, magpies, and brown-headed cowbirds. The starlings compete with songbirds for food, and the cowbirds are nest thieves. When cowbird populations are big, they keep warbler populations down because they take over the nests.

"In past years, the brown-headed cowbirds weren't a problem for warblers because they followed the bison," Lindahl, a long-time bird watcher and former president of the Bridgerland Audubon Society, observes. "But with bison gone, they now concentrate on feedlots, and their numbers have grown. They are extending their range and now are coming in contact with songbirds that have no historical experience with them, so they can't pick out the egg and throw it away, like the ones in the Great Plains can."

But things are looking up where damaging agricultural practices are being curtailed, either by individual farmers or different management practices on the river like PacifiCorp's approach. "I think people are slowly taking a little bit better care of the banks," Lindahl says, "so there are probably more bulrushes. People are planting willows. Those things are on the increase. It's good for songbirds that are right on the edge. We also seem to have more wading birds nesting here. They came up when the Great Salt Lake flooded, and a lot of them that had never nested here before, like the white-faced ibis, went back when the lake level went down again. But a colony still stayed here. So we have a thriving colony of white-faced ibis. There's also a thriving colony of Franklin gulls, and there's one of cattle egrets and snowy egrets."

On the other hand, great blue herons are in trouble because Cache County has so few roosting trees. There are two heron rookeries

Continued on next page

then drying out at inappropriate times as if they were arbitrarily relabeled tide-pool creatures.

At Grace, most of the Bear River is directed into an aqueduct. Consequently, the ecology of the Black Canyon below that point is unique. What little water flows in the river comes from a trickle released from the Grace Dam but mostly from springs and seeps along the sides of the canyon or in the river bottom. Springs that gush from clefts in the basalt walls are rich in nutrients and insects. You can put your hands in them and pull out plants and amphipods (scuds), which thrive in the spring water high in calcium carbonate.

The water in the river through Black Canyon is crystal clear and supports a population of planted and carryover rainbow and brown trout. The water continues to rise as groundwater seeps in until the river reaches the Grace power plant. The gradient through Black in Cache Valley, but some years, they are abandoned. Then a fire about twenty years ago caused some of the few remaining trees to topple over. Dwindling fish populations in the river and its oxbows have diminished their food sources.

Other birds, like the yellow-headed blackbird, are declining over their whole range in the United States but are actually doing quite well in Utah. "We have a large population, and it's stable," Lindahl says. "So there are some things that Cache Valley provides good habitat for, those things that like hardstem bulrush marshes."

The yellow-headed blackbirds don't primarily inhabit the Cutler marsh. They prefer the oxbows and ponds. The marsh is too open for them, although they roost there at night. The oxbows of the Bear riverine system are more productive habitat.

"One of the reasons the oxbows are so productive is because it's impossible to drain them, and the farmers aren't cultivating them as intensively," partly due to Army Corps of Engineer regulations that prevent the loss of wetlands. "It's the major wildlife habitat worth hanging onto," Lindahl observes.

Part of the mountain-brush community; scrub oak at 5,500 feet in Oneida Narrows provides cover for a wild turkey population.

Canyon is low, and the water flows gently and steadily throughout the year, although that will likely change with new hydropower licensing agreements. The slow movement of the clear water encourages the growth of filamentous algae and macrophytes because fine materials collect and provide the foundation for rooting. Without spring floods, the algae and plants are protected from erosion. Equally important, the clear waters foster the deep penetration of sunlight, allowing the plants to photosynthesize and grow, especially in the summer.

From the Cove power plant until the river enters Oneida Reservoir, the land is dominated by agriculture. Once it leaves Oneida Narrows as a clear river, the Bear continues coursing through farmland. When it enters the relict area of Lake Bonneville, however, the river becomes much more turbid, and the water temperatures begin to fluctuate from thirty-two to eighty-eight degrees. In a 1967 survey, Arnold Bangerter noted the paucity of insect life once silt begins to enter the river from the three tributaries upstream from Preston, Idaho. He found only chironomidae (midge) larvae and annelids (worms). Besides flourishing in silt, these species are more tolerant of pollution from farm runoff.

Bangerter found a few mayfly and caddisfly larvae near bridges where gravel and rubble had been imported for improvements but not enough to support a game fishery. Fish sampling in that thirty-nine-mile stretch showed carp as the most abundant fish, followed by black bullheads, green sunfish, suckers, yellow perch, Rocky Mountain whitefish, largemouth bass, black crappies, and a few walleyes.

The floodplain lands along the Bear River in Cache County, Utah, are known as the Bear River Bottoms. Approximately thirteen

Piñon pine and junipers intersperse along the riverbanks in lower Black Canyon, while filamentous algae and macrophytes thrive in the clear, slow-moving, spring-fed water.

hundred acres of these lands are owned by PacifiCorp, the parent company of Utah Power and Light (UP&L), and leased to the Utah Division of Wildlife Resources (DWR). The partnership aims to improve the habitat and try to restore it to a more natural condition. Although there are few historical descriptions, the bottoms most likely supported a cottonwoods/willows habitat, with bushes like hawthorn and chokecherry and marsh/wetland plants like cattails and bulrushes, according to land-management consultants Don McIvor and George Wilson. But once the floodplain became private, many upland sites were cleared, bermed, leveled, and farmed. Some wetlands were drained for grazing and farming by digging ditches and connecting them so that the water drained into the Bear River. Still, wetlands in Cache Valley are extensive and important to the ecosystem.

The Bear River Bottoms play an important role in the life cycle of birds. Many use this area as a stopover, as breeding habitat, or as part of a winter range, although the bottoms lack sufficient food sources to be the sole support for avian species. Nevertheless, the Bear River and its riparian corridor are critically important for birds in

Cottonwoods and willows cluster along the riparian corridor in lower-elevation bottomlands consisting of dense silts and clays.

Sago pondweed in freshwater ponds next to the Bear River provides food for migrating waterfowl.

an arid state like Utah. Pelicans, ibises, and snowy and cattle egrets flourish in the bottoms. The river is a migration route for neotropical birds and provides resting and foraging areas. In fact, Partners in Flight reported the greatest songbird diversity in Utah at a banding station near the bottoms. That same habitat nurtures upland game birds like pheasants and two species of grouse. Deer frequent the bottoms, as do predator birds like great blue herons and ospreys and, occasionally, a bald eagle. Canada geese nest in the marshes, and sandhill cranes use adjacent fields as staging areas. Minks, red foxes, raccoons, and skunks sneak through the undergrowth looking for small mammals, reptiles, amphibians, and eggs.

Upland areas near the bottoms are mostly monotypes. There are some shrubs, like hawthorns, sagebrush, and rabbitbrush, but there are few seed resources for forage. In midsummer, the grasses tend to tangle and fall over, creating an impenetrable habitat for most species.

The bluffs above the floodplain are largely devoid of trees due to clearing. Historically those trees probably constituted a riverine forest with multilayered canopies that supported numerous species of wildlife. Where trees remain, they tend to be far apart, and there are few shrubs. So connectivity is a problem, and the few stands of trees have large amount of edge habitat, sometimes too much, making birds that live there more vulnerable to predation and nest takeovers.

Downstream from the protected bottoms, farmers cultivate almost to the edge of the river, except for the lands around Cutler Reservoir that are owned and managed by PacifiCorp to support wildlife habitat. Farming to the riverbanks has led to less biodiversity.

Below Cutler Reservoir, the Bear continues on its way through agricultural lands until it reaches the Bear River Migratory Bird Refuge. Over that sixty-two-mile course, the river barely seems to move at all. Moreover, a lot of water is taken out at Cutler Dam for two irrigation canals in Box Elder County, Utah: the West Side and the East Side. During the summer irrigation season, the river only receives seepage below the dam and varies from a hundred to twenty feet wide.

Usually the river velocity is too slow for fish to survive, although channel catfish do reasonably well, eating whatever flesh they can find: mice, turkey parts from processing plants, and dead calves, horses, and pigs that are thrown into the river.

Once the Bear enters the bird refuge, it also reaches the last of its vegetative zones: desert shrub. These plants live on mudflats, essentially wetlands with no standing water. The mudflats fill during moist periods of the year but are dry the rest of the time. The ones on the periphery of Great Salt Lake and the Bear River delta require plants that are salt tolerant, like greasewood, salt grass, and pickleweed. Waterfowl often build their nests in salt grass. Mudflats are important habitat, too, for shorebirds like American avocets, black-necked stilts, and sandpipers. Canada geese use mudflats for loafing.

Cattails and hardstem bulrushes in the Bear River Bottoms of Cache Valley support a population of neotropical songbirds, waterfowl, rodents, and small, predatory mammals.

Inside the periphery of the desert-shrub, mudflat community are the Bear River Migratory Bird Refuge wetlands. They are an oasis in the harsh desert of the Basin and Range geography of dry playa lakebeds for more than 260 species of birds that use the refuge sometime during the year. Part of the Western Hemisphere Shorebird Reserve Network, these wetlands are visited by approximately thirty-three species of shorebirds. They are a portion of the two to five million individual birds that rely on the Great Salt Lake annually. Approximately 30 percent of migratory waterfowl in the Pacific Flyway use the Bear River Migratory Bird Refuge as one of their refueling and resting tops, as reported in Box Elder County's Wetlands Management Plan.

The Army Corps of Engineers says that any one of three characteristics defines a wetland. It must have a hydrology where it is inundated temporarily or permanently some portion of the year. It has hydrophytic vegetation, meaning it supports water-loving plants like cattails, rushes, or sedges. Lastly, it contains undrained, hydric (wet) soil that is anaerobic (lacking oxygen) in its upper stratum.

Such wet areas are biologically diverse ecosystems. Most species

Although dry most of the year, mudflats in the delta area of the Bear River at Great Salt Lake are part of the desert-shrub life zone, marked by pickleweed.

of birds utilize wetlands, with approximately one-third of all North American types relying on this habitat for their existence. In the United States, 28 percent of endangered plants and 50 percent of endangered animals spend part of their life cycles in wetlands.

Wetlands remove chemicals, sediments, and excess nutrients from runoff, intercepting polluted waters from point and nonpoint sources before they reach aquatic environments. Wetlands recycle nutrients, and their vegetation absorbs excess nutrients, thus preventing algae blooms in rivers. Suspended solids settle into wetlands and are absorbed by the soils, an important service since 70 percent of pesticides and 94 percent of sediment load can be removed from runoff by wetland vegetation. Wetlands serve as sponges, too, storing water during floods and releasing it during dry periods, supporting base flows to a river and maintaining aquatic habitat. Eventually the stored water in wetlands works its way into aquifers, recharging them.

When wetlands appear in the desert, the way they do around the Bear River as it works its way through its delta into Great Salt Lake, they become the scales upon which all life balances. The wetlands in Box Elder County are a dynamic system providing a rich mosaic of emergent marshes, wet meadows, artesian springs, and saline playas that, when combined with the mudflats, uplands, and open water, create breathtaking abundance and life-sustaining biodiversity. By any measurement, the Bear River delta in the Bear River Migratory Bird Refuge is a critical part of the ecosystem of Great Salt Lake, but it relies on a steady supply of water from the Bear River. If that dries up, the ecology of the Great Salt Lake and the animals that live at its edges will be at risk. An environmental crisis would work its way upstream and spread inland beyond the margins of the lake—toward the population centers of the Wasatch Front.

Finding an Old Fish in New Places

The Bear River and Bonneville Cutthroat Trout

In the summer of 1999, Jeffrey Kershner and his research crew caught a Bonneville cutthroat trout in Thomas Fork, a tributary of the Bear River in Idaho, which startled them for a reason other than its large size. It had a pelvic clip.

"We looked at this fish and said, 'Who's doing pelvic clips!?' because we didn't think anyone was doing it. We started going through all these agencies, and the only people who were doing any pelvic clipping at all was the Utah Division of Wildlife in Bear Lake. It turned out that this was one of their fish. It made it through the marsh there, had come out of the lake, probably the big arm, and into the [Bear] river all the way up forty miles into the Thomas Fork."

Kershner, an aquatic ecologist for the U.S.

Continued on next page

Resident Bonneville cutthroat trout spend their lives in the high-elevation tributaries of the Bear River Basin like Left Hand Fork of the Bear River.

Forest Service and associate research professor in the Department of Fisheries at Utah State University, has been studying the decline of the Bonneville cutthroat, Utah's state fish. His research is part of a larger effort trying to determine why the salmonid's numbers are falling.

At one point, the Bonneville cutthroat was found in up to 90 percent of the lakes, rivers, and streams of the Bonneville Basin. Even after Lake Bonneville shrank to what is now Great Salt Lake, Bonneville cutthroats still were plentiful throughout the relict area. Some anecdotal accounts from early surveyors reported that they took large trout out of the river around Fort Bridger, close to Evanston, Wyoming. They talked about catching "numbers" of fish larger than two pounds and using them to feed their group.

There is evidence now that, when the Bear River flowed into the Snake, there was a connection between the Snake River Yellowstone cutthroat group and the Bonneville group and they evolved together. When the river turned south, the Bonneville cutthroat began to form its own identity. The Bonneville and the Yellowstone cutthroats are more closely related, via DNA, than Kershner and others once thought. But that's just on the Bear. The Bonneville cutthroats in the Deep Creek Mountains in Utah are more isolated and very different from the ones in the Bear River Basin. Yet all cutthroats in the West, whether it's strains of the Bonneville in the Bear River Basin or Deep Creek Mountains or the Lahontan in Nevada, are facing a variety of pressures and reduced numbers.

In February 2005, the Center for Biological Diversity, the Pacific Rivers Council, and the Biodiversity Conservation Alliance filed suit against the FWS for failing to protect the fish as an endangered or threatened species. The suit seeks to force the FWS to list the Bonneville cutthroat under the provisions of the Endangered Species Act. Groups have been attempting to list the species since the 1970s, and the FWS twice determined it merited protection, then bumped it to address higher priorities. When a third petition was filed in 1998, and then rejected in 2001 by the FWS, the coalition sued, feeling that if it did nothing, the Bonneville would perish in spite of the conservation agreement under which Utah, Idaho, and Wyoming agreed to join the Forest Service and BLM in making habitat improvements.

Currently only remnant populations of the Bonneville cutthroat remain in the Bear River Basin, mostly in the headwaters area, and even then, only in the tributaries, not the lakes, which are planted with Yellowstone cutthroats or brook trout. But the Bonneville is proving to be an adaptive fish, according to findings of Kershner and his team. They've been discovering Bonneville cutthroats in the Bear River Basin, where they hadn't expected to find them.

"There are two life-history patterns that are appropriate to talk about," Kershner says, "the first being that Bonnevilles exhibited a couple of different strategies to deal with their environments. There are fish that we call *resident fish,* and those fish are typically in the headwaters," living in the same streams where they were born.

"There's another component to the population that we've just recently started to find again," Kershner continues, "and those are called *fluvial fish.* In other words, they spawn in these headwaters, and they move down into the main river system and spend more of their time in there. In fact, the fluvial fish are typically much bigger than the resident fish. We find fluvial fish up around the Thomas Fork of the Bear and the Smith's Fork that sometimes are well into twenty inches or more."

The fish spawn in the tributaries but move into the main stem of the Bear at various times of the year. Kershner and his team have found the fluvial fish throughout the Bear from above Bear Lake downstream to Alexander Reservoir at Soda Springs, Idaho.

"We don't see them during the hot summer months when the river is dewatered and things are tough," Kershner observes. "But we see them showing up again right around the end of October and November. Our speculation is that they're spending the critical summertime up in the higher elevations where they can actually connect and get back to cold-water habitat, and they spend their time after spawning up there. Then, as the temperatures drop, they move back into the river—if they have a chance."

Kershner theorizes that the fish move down into the main stem partly because the main river doesn't freeze solid and ice formation is less of a problem than in the tributaries. Equally importantly, the main river runs clearer in the fall and winter, and increased visibility facilitates predation.

"The one thing we do see with these bigger fish is, bigger fish eat fish," according to Kershner. "Some of these tributary streams [like Thomas Fork] have nongame fish like carp. We've pulled small carp out of their [Bonneville] stomachs. Obviously they've found a food source they like. Almost all of the big fish we've sampled have some fish in them. So they probably have better opportunity to get those nonnatives and do a little more predation down there [in the main stem]."

Ironically, a foreign fish introduced into western waters as a possible food source for human beings has become a gourmet menu item that may help the Bonneville cutthroat, a native fish, to eke its way back to healthy numbers. That is, if it weren't for other threats on the Bear. Kershner is quick to add that the fluvial Bonneville he finds are few and far between.

"What the dams have done is truncated those distributions by cutting off habitat, by

Continued on next page

Nymphs feed on microorganisms breaking down the decomposing body of a cutthroat trout in the shallow Allsop Lake inlet.

cutting off tributary space," Kershner says. "So we don't have what we call 'connectivity' between the populations prior to all these diversions."

Another threat to the Bonneville cutthroats is hybridization with other trout, especially rainbow because they are so closely linked. Eventually the planted rainbows and rainbow/cutthroat hybrids begin to push out the pure Bonneville.

"Certainly hybridization is an issue for cutthroat, particularly where you've got introduced rainbows in that Oneida area and all through the river," Kershner says. "In fact, we've done work on the South Fork of the Snake, where we've got continuing concern. Rainbows were introduced there all the way up to the mid-'80s, and then all of a sudden, the rainbows started to become a bigger proportion of the population. They've continued to grow since around probably 1987. The number of hybrids in there are starting to increase as well. So we know that's a problem."

Can hybridization be a reason for listing a species?

"Yeah," Kershner replies. "We had, in fact, a court decision that was just rendered for West Slope cutthroat in Montana that says, given this possibility of hybridization with nonnative species, that you need to go back and reexamine the petition and look again. So yeah, there's some potential there."

The incidence of Bonneville cutthroats in the Bear River system poses real problems for PacifiCorp, whose dams truncate the movement of the fish, when the time comes to relicense hydropower generation permits, like the recently completed permit for the utility's Soda Point,Grace,and Cove operations. When Kershner and his research team began finding fluvial Bonneville in unexpected stretches of the river, he reported that fact to the interagency group that was studying PacifiCorp's licensing request.

"In about 2001, after we had done some preliminary analysis of the data and a report for the Fish and Wildlife Service, we dropped this bomb on the relicensing group and coordinating committee. In fact, the interagency group that is part of the relicensing process has gone back and said, 'Now we're really interested either in some mitigation or trying to figure out what we should do in the future.' One of the proposals submitted by Trout Limited was to try and reconnect some of that by getting rid of that small dam that's at the bottom of the Cove bypass and then opening that piece up to some of the tributaries there."

It will require political muscle to get PacifiCorp to agree to the mitigation idea of bringing back the Bonneville in areas below the Grace power plant. To provide access to spring-fed creeks coming into the river downstream, PacifiCorp may have to give up power generation at Cove or figure out another way to divert through a re-engineered concrete flume or aqueduct. If PacifiCorp agrees there is a need to rehabilitate the Bonneville in that stretch of the river, other tributaries in Gem Valley, Idaho, like Williams Creek, offer excellent spawning habitat for the Bonneville.

In fact, several stretches of the Bear River could become blue-ribbon, Class-I, coldwater fisheries. Right now these classifications only are given to a few tributaries, like Blacksmith Fork and the Logan River in Cache County, Utah. The spring-fed stretch through Black Canyon has that possibility, especially if increased flows to the river don't add to the turbidity. The stretch from Oneida Dam to Riverdale, Idaho, shows promise, too.

"When you get out of Oneida and you head down toward the hot springs [Riverdale]," Kershner explains, "that stretch of river that's between where the river comes out of the canyon [Oneida] and then makes the big wide turn to head toward Utah, that stretch of the river... I've actually walked in the fall and have seen brown trout spawning around the islands. You can actually see redds. Now whether there are cutthroats down there that actually spawn in that section and use places like Mink Creek, I don't know. That's part of the thing that I was proposing to Idaho Fish and Game. But we really do need to take a different look, go out there and take a look at different times of the year to see what's going on."

Water seeps through fissures in the volcanic rock walls of Black Canyon and life takes root.

You give me a name,
Ursus,
To carve your sign
On virgin land.

Or so you believe,
Righteously claiming scant bones
Mean nothing in your way.

I mark the path
To the garden.

You spill life
On my banks.

You stop,
Looking forward,

And ask that I water your roots.

The Stage Coach Inn at Hampton Ford served three different stage companies. Now it serves a variety of new community needs, including receptions.

Chapter 4

The Bear River and the Threads of Western American History

Many of the great themes of western American history were stitched together on the banks of the Bear River. Native Americans and Anglos antagonized each other, and those abrasions erupted into some of a tense nation's most horrific, soul-searching moments. European mercantilism and colonialism squared off against an emerging, entrepreneurial spirit some were beginning to call "American." Financiers salivated over untold natural resources and sought to bring them to burgeoning cities in the East. Developers promoted western expansion and the benefits of land and climate in pamphlets that were sometimes quaint and other times flat-out wrong. Settlers and vagabonds never ceased dreaming of a second chance.

Native Americans

The first inhabitants of the Bear River Basin began appearing 12,000 to 15,000 years ago as the last Ice Age ebbed and the western climate warmed. Those Desert Archaic or Western Archaic cultures, largely nomadic tribes that exploited seasonal environments around Great Salt Lake and the Great Basin, lived in wickiups of interlaced brush and sticks in the summer and larger underground pits with roofs of logs and organic material in winter, often with multiple families. One theory suggests that these groups evolved into the later Fremont culture.

Steven Simms, an archaeologist at Utah State University, looked at the Fremont culture when his research team did a reconnaissance of the Bear River marshes around Great Salt Lake in 1990. They determined that the area had been occupied for thousands of years, although most sites they found were approximately fifteen hundred years old. The older settlement areas have been buried under at least one meter of sedimentation from the river as it braided from side to side in its delta.

During the Fremont and Late Prehistoric period that extended from the fourth to the fourteenth centuries, the Fremont peoples used the lower Bear River and associated wetlands as farming bases, as well as a staging area for typical hunter-gatherer activities. Great Salt Lake provided them with waterfowl, fish, small and large mammals, and numerous plants. Equally importantly, the marshes offered access to uplands for shelter and more food. During the periods when Great Salt Lake rose, the marshes next to the lake became full of brackish water, limiting the food supply. Then the Fremont moved to the wetlands along the Bear River and its tributaries and springs to exploit those environmental niches.

There are two schools of thought about how the Bear River wetlands affected Fremont lifestyles. Wetlands could have increased diversity in ways that the Fremont organized themselves. Some could have opted out of agriculture altogether, focusing instead on hunting and gathering around the wetlands. Or the wetlands could have supplemented individual agricultural practices, increasing the variety and abundance of food. Each of these options would have led to different combinations affecting mobility and group size.

During the fourteenth century, the Fremont culture gave way to modern Indian tribes, the ones who populated the Bear River Basin once histories were written. While the Utes ranged north to Utah Lake and the southern shores of Great Salt Lake, the largest tribes in the basin were the Shoshones and then the Bannocks. Both belong to the Shoshonean linguistic branch of the Uto-Aztecan family and were primarily nomadic hunter-gatherers.

The problem is that Anglos just entering the region wrote many of the contemporary descriptions of Indian tribes, and they knew little about the sophisticated lifestyles of different Indian populations. They tended to lump all Indians together, not recognizing historical alliances or enmities, or they named them according to what they saw them eat on one occasion, not realizing that Indian foodstuffs changed with the seasons.

For instance, early Anglo journals refer to Indians called "the Snakes." For some diarists, that meant Indians who lived in the Snake River drainage. For others, it meant any Indian they encountered between South Pass and the Columbia River. The designation of the Snakes came from the French term *gens du serpent* or *les serpentes* from the sign language for these people. Generally, however, the term referred to the Shoshones.

Anglos tried to make a crisp distinction, too, between the Shoshones and Bannocks, who often traveled together and intermarried. They described the Shoshones as friendly to whites and the Bannocks unfriendly and warlike, when in fact the cultural traditions of the Bannocks were similar to the Shoshones. The Shoshones, Bannocks, and their linguistic relatives to the west and north simply called themselves "the people."

Confounding the Anglo categorization and cultural descriptions of the Shoshones, especially, was the fact that there were two distinct lifestyles within the Shoshone culture, based on whether the tribes were horse mounted or foot bound. Those who relied on horses ranged more widely in their hunting and gathering and participated in buffalo hunts in the northern sections of tribal territory. These Shoshones often banded together in larger groups to hunt big game more effectively. Their social organizations were different, too, with headmen picked for their hunting prowess, although seldom were there permanent political/social units and chiefs. The foot-bound Shoshones, on the other hand, had a different set of food choices. They also were seasonal and reflected the Indians' keen understanding of their environments and what they provided since they traveled widely to hunt and gather, albeit on foot. Their social organizations tended to be nuclear families or a few related households since their available food wouldn't accommodate the larger groups formed by horse-mounted Shoshones.

In his 1845 report to the U.S. Senate on his western exploration, John C. Fremont commented on the emaciated Indians he found fishing in the wetlands below what is now called Cutler Narrows. Calling them "Root Digger" Indians, he noted they made their nets from canes and rushes. But their appearance was shocking: "They had unusually large heads with matted hair and [were] almost entirely naked." In another passage, Fremont described a group of Gem Valley Indians that he came across in August. He noted that they were "miserably poor," and their "subsistence is almost solely derived from roots and seeds, and such small animals as chance and great good fortune sometimes brings within their reach. … Roots, seeds, and grass, every vegetable that affords any nourishment, and every living animal thing, insect or worm, they eat."

Fremont's descriptions were ethnocentric at best and reflected his Euro-American biases for certain foods. The foot-bound Shoshones had a varied diet of nutritious foodstuffs that gave them choices. At different times of the year, they ate yampa and camas roots, wild rye seeds, gooseberries, currants, buffalo berries, and chokecherries. In the fall, they migrated to areas around Great Salt Lake and feasted on protein- and fat-rich pine nuts. At other times, they banded together with other Shoshones to catch hares in nets. In some ways, they had a more varied and nutritious diet than Fremont and his men. As a culture, they had adapted to a feast-and-famine regime. They put on weight during seasons when food was plentiful and stored foodstuffs for winter, when they lost that weight to generate body heat when supplemental food was hard to find.

So, when Fremont happened to see a "miserably poor" Shoshone family in August 1845, he generalized too much. It could have been a single family who had recently been preyed upon by another tribe; perhaps they had had their horses taken by Blackfoot Indians, who often entered Shoshone territory to steal horses. Perhaps they were a family recovering from one of life's misfortunes, as Anglo families sometimes have to do when facing sickness or death. The point is that Fremont and other early western explorers/observers, due to their cultural biases, mislabeled Indians and often used condescending descriptors that bordered on racial stereotyping. Unfortunately, that language tended to dehumanize the Indians of the Bear River Basin. The resulting mental set it generated in Anglos made later inhumane treatment of Indians more acceptable and "culturally correct."

Site of the January 29, 1863, Bear River Massacre, where approximately 250 Shoshone men, women, and children were slaughtered by forces of Col. Patrick Edward Connor from Fort Douglas, Utah.

The Bear River Massacre

With the Civil War as political backdrop, Anglo and Native American tensions in the 1860s in the Intermountain West had reached the flashpoint. It would take cool heads to avoid a horrific confrontation. Sadly, there were none to be found.

While the Shoshones originally had been receptive to Mormon emigration, they became angry over the results. Within fifteen years, the Northwestern Band of the Shoshone Nation had seen a relentless encroachment onto their aboriginal lands, upsetting their traditional hunter-gatherer subsistence, according to Brigham Madsen in his pivotal history *The Shoshoni Frontier and the Bear River Massacre*. Grasses used to grow knee-high before settlers arrived, and the Shoshones harvested the seeds for a nutritious mush and winter storage. But when settlers let loose great herds of cattle onto these fragile pastures, the grasses quickly were chomped to the ground. Increasingly there were no grass seeds to harvest.

Also the Shoshones had to compete with white settlers for wild game in the mountains and marshlands around the Bear River. Because the Anglo population was growing faster than the Indian one, wild-game numbers were not increasing, and hunting was not managed, large numbers of deer, rabbits, geese, and ducks were decimated. The Shoshones were facing starvation.

Originally Brigham Young promoted a policy of "it's better to feed them than fight them," and Mormons tried to be generous in their gifts of food. But as Mormon numbers grew and traditional food sources began to shrink at an alarming rate for the Shoshones, the Mormon settlers couldn't meet the demands of the Indians. Resentment began building on both sides. Indians began to make periodic raids on farms to demand or steal food. Whites retaliated.

One of the problems exacerbating Anglo-Shoshone relations was that no formal treaties had yet been struck, so there were no established yearly payments in exchange for a formal delineation of land boundaries. Mormons had long coveted Cache Valley because of its water supplies, but that was also one of the principal, if seasonal, residences of the Shoshones. In 1855, when the Utah territorial legislature granted ownership of Cache Valley to Brigham Young as trustee for the Church of Jesus Christ of Latter-day Saints as a way to formalize Mormon claims on the land, the Shoshones received no restitution. Nor was there a legal avenue to redress their grievances.

By the 1860s, emigration along the Oregon and California Trails had reached its zenith. Oxen and cattle had denuded the grasses along the trail, further worsening the predicament of the Shoshones. The establishment of overland mail and freight routes added to the problem, and Indians—and whites posing as Indians—attacked mail carriers, traveling miners, and wagon trains, stealing livestock. Whites demanded that the authorities protect them and their property. Sometimes emigrants fired deadly potshots at Indians along the trail, regardless of whether or not they were threatening.

Federal authorities were no help. Indian agents turned over at a rapid rate, and there was no stable policy toward Native Americans. Moreover, Congress wouldn't grant the Office of Indian Affairs and its superintendents enough financial resources to allow the agents to intervene and feed starving Indians. Mistrust also existed between federal officials and Mormon leaders. This was just after the era of the Utah Expedition, known as Johnston's Army by Mormons, sent to enforce federal authority, and a legacy of suspicion permeated Indian affairs. Mormons didn't like the federal presence, and government authorities had reason to believe that Mormons were manipulating Indians to bolster their territorial control.

Enter an army unit commanded by Colonel Patrick Edward Connor. Connor was charged by the Department of the Army to raise a regiment to protect the mail line. In July 1862, his Third California Volunteers left Stockton on their march to Salt Lake City. Connor had failed at several ventures in California, but he was acknowledged to be an energetic, shrewd leader and soldier. He hungered for promotion to brigadier general. The volunteers, unhappy that they hadn't been called to the East to fight for the Union in the Civil War, wanted to distinguish themselves. Indians would have to do.

Gold was discovered in Montana in 1862. That meant that, in addition to east-west emigrant traffic through Shoshone territory, there now was a north-south incursion by prospectors and freighters hauling supplies on the Montana Trail. Then, in December 1862, Major Edward McGarry from Connor's regiment hung four Indian hostages at Empey's Ferry and threw their bodies into the Bear River in retaliation for horse theft. Indiscriminate killing on both sides accelerated, and a kind of frontier ruthlessness set in as whites felt that Indians needed to be taught a lesson once and for all.

Intelligence reported that about 450 Shoshones were encamped at the confluence of Battle Creek and the Bear River a few miles upstream from what is now Preston, Idaho. The opportunistic Colonel Connor wanted to

Continued on next page

After the Bear River Massacre, many of the Shoshones converted to Mormonism and settled at Washakie in northern Utah. Other Shoshones went north to the reservation at Fort Hall, Idaho.

move quickly before the Indians went elsewhere. In mid-January 1863, Connor began secretly sending north small detachments of riflemen and cavalrymen. He hired Orrin Porter Rockwell to be his scout. The army set out with cannons, but they got stuck in snowdrifts. Seventy-five men had to pull out due to frostbite, so that by the eve of January 28, Colonel Connor had a force of about two hundred men well fortified with ammunition. They met a force of slightly less than two hundred Shoshone warriors, who had much less ammunition and were encamped with many more family members.

On January 29, 1863, Colonel Connor's forces attacked. The fight lasted four hours. Eventually the Indians were flanked, and wholesale slaughter began. Leadership broke down, and officers fought side by side with enlisted men in close quarters, killing Indians with deadly revolvers. A "smoking gun" diary entry from a participating soldier found more than a hundred years later indicated that Connor gave the order to take no prisoners. Soldiers began to kill women, children, and the elderly. The wounded and incapacitated were struck in the head with axes. The killing continued even after some Shoshones attempted to surrender. When it was over, there were estimates of 250 to 350 Shoshone men, women, and children killed; Madsen feels that 250 is a reliable number. Even using this lower figure, this encounter marked the largest massacre of Indians in western U.S. history. Settlers said Connor's success showed "the intervention of the Almighty."

Commemoration

On January 29, 2002, approximately fifteen people met at the Bear River Massacre site to remember what had happened. They had been meeting since 1999, led by the Northwestern Band of the Shoshone Nation and the Friends of the Native Americans of Northern Utah. They wanted to honor the dead, as well as become a visible lobbying force demanding the return of sacred ground to the Shoshones.

Leland Pubigee, vice chairman of the Northwestern Band of the Shoshone Nation, began the commemoration ceremony by burning cedar bark "to cleanse the place." At the end of the ceremony, he burned sweetgrass "to sweeten the atmosphere," although he noted that the once-plentiful grass was harder to find now.

Patrick Mahoney, a reenactor from the Third California Drum and Fife Corps from Fort Douglas in Salt Lake City, solemnly

A participant at the 2002 commemoration ceremony hangs feathers on a tripod to protect and remember the Shoshone spirits who wander the area.

represented Connor's Third Volunteers. He used to come to the commemoration in his period uniform until he realized how the symbol continued to wound the Shoshones. Since then, he has dressed in civilian clothes out of respect and brought ceremonial gifts for the tribe, representing the four cardinal points of the compass.

Kerry Brinkerhoff, president of the Friends of the Native Americans of Northern Utah, was more fire and brimstone in his comments. He reported on the frustrations of trying to find a champion in Congress to establish the grounds as a National Historic Site and allocate funds to purchase the land. Gesturing across the highway to a pasture where the Shoshone tribe had nearly been wiped out, Brinkerhoff asked the rhetorical question, "Would this land be filled with cows if this were the site of a massacre of Mormon pioneers?" Brinkerhoff said that the lack of interest from Intermountain senators reflects the ongoing marginalization and lack of respect for Native Americans and, especially, their massacred dead.

Patrick Mahoney began the healing part of the ceremony, hanging a blue and a white feather on one of the lodgepole tripods. The feathers were meant to protect and remember the spirits who wandered the area.

After the commemoration, some of the congregation wandered away from the day's ceremonial site with its wooden, declarative Bear River Massacre sign to another, stone monument. This memorial testifies to the difficulty that people have coming to grips with their past and reveals the tensions that arise when

Continued on next page

historical perceptions have to wrestle with reality.

The monument has four sides, with metal plaques on three of them. On the east side, the Franklin County chapter of the Daughters of the Utah Pioneers (DUP) erected a plaque in 1932 that calls the encounter "the Battle of Bear River" and says that the Shoshone women and children were "combatants." On the west side, the DUP has another plaque entitled "Pioneer Women." The gist of that plaque is that settlers in the area had been plagued by Indian attacks and the troops from Fort Douglas had taken care of the problem once and for all. The plaque also notes that the pioneer women in the area, used to travail, took the injured soldiers into their homes, dressed their wounds, and nursed the most severely hurt back to health. There is no mention of the dying and wounded Indians and what happened to their bodies, which were left to the wolves and coyotes. On the north side, however, the National Park Service in 1990 erected a simple metal plaque that labels the memorial the site of the Bear River Massacre. The fourth side is reserved for the Northwestern Band of Shoshones to place their side of the story, but they refuse to dignify the divergent viewpoints with their participation. They know the battle was a massacre, pure and simple.

It has been difficult for the Shoshones to mount an effort to truly commemorate the area because the massacre site is now owned by fifty different entities. But in 2003, the coalition finally got someone to listen. Although Congress still has not followed the National Park Service's recommendation to elevate the battleground from a national landmark to an historic site, the Trust for Public Land, with the help of the American West Heritage Center in Wellsville, Utah, purchased twenty-six acres and turned over the deed to the Northwestern Band of the Shoshone Nation. That troubled land next to the Bear River is back where it belongs.

Fur Trappers

The first Anglos to enter the Bear River Basin at Gem Valley came from the north via the Portneuf River in 1811. They were a group of five trappers nicknamed "the Kentuckians" and were part of John Jacob Astor's American Fur Company. Although they didn't keep a journal and some of them were killed by Indians along the Boise River a few years later, they did report their explorations to Robert Stuart. He was intrigued and decided to see if the river they described would be good for trapping beaver.

Stuart was impressed with the broad plain he described in his journal. The only problem was that the grass had been burned by the Indians, either as a hunting technique or to release carbon into the soil for faster growth. Stuart named the large river Miller's River after one of the Kentuckians, Joseph Miller. But the name didn't stick.

William Clark saw the name Miller's River on a map published in *Nouvelles Annales*. But he decided to call it Crooks River because a man named Crooks from Stuart's party had delivered the map. Alexander Ross, writing a history of the North West Company, called it the Sherrydikas, the name he gave to the band of Shoshones called Sherrydikas or Dog Eaters. Later, William Ashley reported in 1825 that it was the mythic Buenaventura River, the long-sought, but mostly dreamed about, outlet from the inland West to the Pacific Ocean.

The name that eventually stuck came via William Kittson, a clerk in Peter Skene Ogden's trapping brigade. He called it the Bear River, remembering the name that Michel Bourdon, a fur trapper from Donald McKenzie's North West Company, who had visited the basin in 1819, had given it. Bourdon felt Bear River was most appropriate due to "the great number of these animals on its borders."

Of course, these mountain men came into the West in search of wildlife with their pelts and meat. Stuart reported "considerable fresh sign of Buffalo" but "at present appears no more the asylum of the ingenious Beaver, than do the bleak summits of the Cordilleras."

The buffalo didn't last long in the Bear River Basin. After six male buffalo were killed by a later Ogden party in 1829, his fur trappers did not note any more in succeeding years. Captain Benjamin Bonneville, who came down the Bear River in 1833, recorded seeing buffalo, but none were reported after that.

In 1845 John C. Fremont reported seeing antelope and elk in Gem Valley, in addition to cranes, geese, ducks, and grouse. Earlier, mountain men had commented on the abundance of mountain sheep at the northern terminus of the Bear River Range near Soda Point. They named the formation Sheep Rock after their successful hunts. But the animal that most caused the pulses of fur trappers to race, and launched a global competition for natural resources, was the beaver. That wetlands mammal represented what gold did to opportunists some twenty-five years later.

The British crown had forced the North West Fur Company into a merger with the Hudson's Bay Company in 1821. The Hudson's Bay Company, in turn, maintained the North West Fur Company's policy of trying to wipe out the beaver in the Snake River area of southeastern Idaho as quickly as possible. They weren't sure how much longer the British would be able to hold onto the Oregon country, and they wanted to dissuade the Americans from moving in, believing they were only interested in trapping beaver. If there were no longer any beaver, they theorized, the Americans wouldn't be interested in staying around and colonizing the area. But the British were at a competitive disadvantage due to their cumbersome organization and antientrepreneurial work rules. And it didn't help that the Hudson's Bay Company was an English corporation that had to practice diplomacy with the Americans, who were more independent operators.

The British crown had granted royal licenses for trapping in the colonies, and British companies employed an *engagé*, who was a wage earner not too far removed from an indentured servant. He often found himself indebted to the company, too, for traps, guns and ammunition, and provisions. The Hudson's Bay Company also employed *freemen,* independent operators not obligated to work for the company but often indebted to it for supplies. The freemen typically were offspring of French trappers and Canadian Indian women. They were allowed to bring along their wives. Ogden's problem with the freemen was that he was limited in what he could pay them for their pelts. When Americans offered the notoriously unreliable freemen a higher price, they often deserted to the other side to Ogden's irritation, especially when they didn't repay the company what they owed.

When Ogden's party first started trapping on the Bear for the Hudson's Bay Company, William Kittson reported taking 133 beaver and one otter on April 27, 1825. They bagged 20 the next day and 16 the day after that. April 30 was a good day for them; they trapped 76 beaver, largely by working the tributaries to the Bear. On May 2, Kittson said they found 74 beaver and one pelican in their traps. Work was interrupted for a few days when one of the party was attacked by a grizzly bear while setting his traps in Cache Valley but saved himself by jumping into the Bear River.

They trapped 319 beaver in just six days. But Ogden was disappointed. He had expected a larger haul. He told Kittson that he felt like a Johnny-come-lately, suspecting that an American trapping party had worked its way down the Bear before him. That had been happening a lot.

When trappers talked about how best to trap beaver, they felt that smaller parties would be more profitable because larger groups tended to spook the animals. Trapping bosses theorized that six men with six traps each trapping during two months in spring and three in fall, for a total of 131 working days, could harvest 4,716 beaver.

By most measures, then, the Hudson's Bay Company brigades were bloated and unable to compete effectively. Ogden's party, for instance, consisted of fifty-nine men, equipped with guns, 352 traps, and 268 horses. Thirty freemen brought their wives, along with their thirty-five children. Logistical time was eaten up, too, when the brigade had to leave the field periodically to return to the company posts for supplies.

John Jacob Astor's American Fur Company was a stiff competitor in the upper Missouri River area. But the primary competition for Ogden's Hudson's Bay Company brigade in the Bear River Basin was the Ashley-Henry Company out of St. Louis. The list of names responding to Ashley's notice in the *Missouri Republican* in 1822, requesting "enterprising young men," would become legends in western American history. Among them were Jim Bridger; the Sublette brothers, Milton and William; and Jedediah Smith, who later became Ashley's partner in the successor Rocky Mountain Fur Company.

Ashley followed the organization that trappers felt was most efficient. Instead of the large brigade the British sometimes used, necessitating periodic departures from the field to resupply, the Ashley-Henry Company sent out teams of seven trappers who stayed in the mountains year-round. These trappers purposely worked the tributaries and areas where they were better protected from Indian attacks. Moreover, they worked in Shoshone and Ute country, and those Indians weren't trappers. In 1824 Jedediah Smith, under the command of William Ashley, took six men and went upstream from Soda Point to the Uinta Mountains. He and Ashley agreed to meet the following year, when Ashley would bring supplies from St. Louis.

The trapping life never was easy for the mountain men. Although they often tried to winter in Gem or Cache Valleys, first meeting in a designated location to spend the winter months together, even these lower-elevation locations could be harsh. In the severe winter of 1825–26, the mountain men had to relocate and winter at the mouth of the Bear River, where they were joined by Shoshone Indians.

Don Mabey reported the men never could kill enough game to keep themselves plentifully supplied with meat since they required an average of nine pounds a day to produce enough calories to do the heavy work and maintain body heat during subzero days. Snow was typically up to the horses' bellies, and they periodically broke through the crust. Moreover, the trappers had to keep their horses herded close together to prevent them from being stolen. Consequently, the horses could find little feed once they had scraped through the snow in the immediate vicinity to find the meager dried grass beneath. Since the Bear River was frozen enough for horses to walk on, the only water they typically drank came from snow and icicles.

At least the American trappers could look forward to warmth, partying, and overall binging in July because of Ashley's smart marketing move of organizing them around the rendezvous system. Instead of forcing the men out of the field to re-supply, Ashley brought needed replenishments and purchased their pelts there. By staying in the field longer, the trappers could harvest more fur. The main rendezvous was in Cache Valley, then known as Willow Valley, in 1826 and 1831, and at Bear Lake, usually called Sweet Lake to differentiate it from Great Salt Lake, in 1827 and 1828. Indians were drawn to the rendezvous, too.

By 1834 the American Fur Company had bought out the Rocky Mountain Fur Company and hired the remnants of Ashley's original men. The cutthroat underpricing and business tactics available to a larger company, coupled with diminishing numbers of beaver, proved effective. The American Fur Company became a monopoly in western fur trapping.

Trapping in the Bear River Basin

A Tradition from the Nineteenth Century Uneasily Confronts the Twenty-first

Stan Bassett reaches into his pickup truck for a string of leg-hold traps.

In 1963, when he was eight years old, Stan Bassett received a gift from his parents that would shape the rest of his life. He had seen an entry in the *World Book Encyclopedia* on trapping and had been bugging his dad unmercifully about getting a number-one long-springed trap. On Christmas morning, it looked odd wrapped with a bow.

"The following spring where I lived in Hyrum," Bassett says, "we had a ditch running behind the house. I went over there and made a picture-perfect set from the encyclopedia. And I caught my first [musk]rat, and boy, was I hooked. From there it's evolved, and I've done it ever since."

Although Bassett has two bachelor's degrees from Utah State in sociology and psychology, he decided to follow another vocation: roofing. It pays the bills for his family of four, but more importantly, because the roofing business slows to a halt when snow begins to fall, it enables him to pursue his passion—trapping—during winter, the most productive time of the year.

"Would you say you're into trapping more for the money, or is there something else about it? Would you do it if you weren't making money?" I ask.

"For the love of the sport," Bassett replies. "I work to make money. When I trap, if I itemize everything up, I probably lose money. I like to do it because you get out, and you learn a lot about wildlife. And anything you get out of it is just a bonus."

Bassett is president of the Utah Trappers Association (UTA), an organization that stresses ethical trapping practices and helps make a market for pelts. The state of Utah regulates trapping through the DWR and oversees such regulations as the interval between when a trap is set and subsequently checked, the price of a license, a system identifying trap ownership, and how and where to use bait. The state also passed a law that people born after 1984 need to pass a class on trapping ethics to get a license. Bassett and the UTA provide these classes to support the letter of the law, as well as preserve and promote the spirit of trapping. In addition, Bassett instructs youngsters on how to study the environment for clues about animal habits, how to attract particular types of game with different traps and sets, and how to preserve and grade pelts for sale.

The UTA meets once a year to create a market where trappers can sell their pelts. Buyers from Europe and Asia mingle with domestic

Larger Conibears are used to trap beaver. Bassett attaches the trap with heavy-gauge wire to a stiff pole driven into the riverbank on his family's property so that the trapped beaver can't swim away.

sellers and set prices for beaver, raccoon, ermine, coyote, mink, skunk, muskrat, bobcat, and fox. When fur prices go up, so does membership since you have to be a member to sell furs at the convention. Currently there are 350 core members.

I ask Bassett to talk about the historical prices for fur, using the muskrat as an example. "When I first started and I was a kid," Bassett begins, "it was a buck and a quarter to $1.60 for a large rat. That wasn't bad. Then in the mid '70s, early '80s, we had a fur boom. I took a couple rats over to Ed Val Carson [a local buyer], and he said, 'Well, I'll give you six bucks a piece for the two hundred, or we can grade them from $8 down to whatever.' And there were a ton of rats then."

"Then the bottom crashed," Bassett continues. "I sold seven hundred rats, and I got $1.10 average on them a few years back. In 2002 they came up a little bit, so at our sale last year, I think they were at a $2.70 or $2.80 average."

Based on inflation, does that mean they are worth less? "Oh yeah, far less," Bassett replies. "And I don't know what's going to happen. The fur market right now is just really volatile. It's a lot like our economy. With our dollar dropping right now, it makes it so the foreign countries can buy more fur. So that's a plus. But with the crisis in the Middle East, that affects stuff."

"And fur isn't probably the kind of thing that people purchase when they are concerned about the economy," I add.

Continued on next page

Bassett harvests a beaver on his family's property the next day. He's disappointed that it is a juvenile kit.

"No, not in America. Fur is considered a luxury item. Most fur you'll see in America is in the trim trade, where you'll see strips of mink or coyote here on the collar. It's not utility use. Whereas in Russia, Korea, and countries like that, they wear it because they need to where it's cold. In the U.S., it's not worn to keep yourself warm. It's more for the look or the feel."

Bassett uses three kinds of traps: a Conibear, which is a square of heavy-gauge steel and spring that is used to trap beaver and raccoons; leg-hold traps, which are used to catch smaller mammals like muskrats, skunks, and minks or when the trapper wants the option of letting a trapped animal go, like a bobcat kitten; and snares, which work effectively on foxes and raccoons. The leg-hold traps cost anywhere from three to eight dollars, while the larger Conibears sell for twenty-five dollars. On an average trapping day, Bassett sets about four hundred dollars worth of traps.

A trapper like Bassett can be successful and make some money during times when a commercial fur farmer may go under. Farm-raising mink or other fur-bearing animals is a risky business because it's capital intensive; at the same time, prices are volatile. For example, a mink rancher is a captive of the market. If it costs him *X* amount of dollars to raise a mink and the market will pay him less than that for the pelt, he goes out of business, although in 2003 prices for farm-raised mink were up. But a small entrepreneur trapper, who doesn't have to sink money into the farming infrastructure, is more flexible and better able to weather price volatility.

"You've got your traps," Bassett explains. "You don't have to feed the animals, and if you do your scouting, you can take a pretty good catch. And they don't raise 'coons domestically, so you have to catch them. Wild mink are different from ranch mink. They all look like mink, but if you put two furs together, the wild mink looks completely different than a ranch mink. And it's hard to raise muskrats domestically, whereas you can have a marsh, and they thrive by themselves."

Trapping in the twenty-first century stirs a lot of public passions. People for the Ethical Treatment of Animals (PETA) are known to throw red paint on fur coats and yell "murderer!" at the wearer. The Fund for Animals has a national program promoting antitrapping legislation in the states. A poll commissioned by the Animal Protection Institute determined that 58 percent of Americans would be "willing to ban all trapping." Yet the Wildlife Society, an international group of wildlife professionals, supports trapping, believing that it is part of the North American heritage and that regulated trapping that supports natural-resource conservation, maintains ecological processes, preserves genetic diversity, and ensures the existence of species and ecosystems.

Bassett senses that there has been some backlash against PETA tactics, too. "That political correctness—those guys kind of hurt themselves," Bassett observes. "I think this is America, and if it's legal, and you have a right to wear fur, and someone comes along and throws paint on you to represent blood, that doesn't wash too well with the majority of the people. Plus what kind of gripes me on that is that an individual that does that say they love the animal, and they respect it so much, yet they would degrade that coat by throwing that on."

Nevertheless, Bassett is sensitive to the arguments of the opposition. He believes that, done correctly, trapping is humane. When set properly, a Conibear trap kills instantly by crushing the animal's neck or backbone. His snares are on a swivel so that wild animals can't entangle themselves and suffer before they are strangled. Bassett uses a more-expensive, stop-loss leg-hold trap with a second spring that immobilizes the animal so that it doesn't twist off its leg. All his leg-hold traps are set near the river's edge and weighted so that an animal quickly drowns. Moreover, when he traps for bobcats, he uses a coil-spring trap with only two coils instead of four so that, although it holds the animal, it only immobilizes it without cutting into the front or hind leg. Then, if it's a pregnant female or kitten, he can set the animal free. Still, Bassett acknowledges that a successful trapping venture means the death of an animal.

Trapping in the Bear River Basin Today

"How many people besides yourself do you think are trapping in the Bear River Basin?" I ask.

"Um, I'd guess a hundred," Bassett says. "And of those, you get a lot that are a weekend-type guy who goes out and sets some out on Friday and pulls them in on Sunday. Your hardcore trappers that I know of—there would be one, two, three, maybe half a dozen that are your legitimate, really go-getters."

Trapping in the Bear River drainage can be fairly successful today. Bassett doesn't think it's dying out, either. "If you put in your time," he claims, "you can catch a lot of fur bearers. *If you put in your time*. If you just go out there and are not really too aggressive at it, you get usually what you put in. In order to catch anything and do it very well, it's just like with anything. You have to preseason scout. You have to get out and make your sets and check them regularly. Keep them maintained really well, and it's a lot of work."

Trappers tend to be a secretive bunch. It's not a team sport. Moreover, trappers worry about "the antis" spying on them, pinpointing where they set their traps, and then springing or taking them, even though it's against the law to interfere with a trap. Bassett wears camouflage when he traipses the streams in Cache County, trying to draw less attention. There's also an ethic among the longtime, devoted trappers not to trap in another person's territory.

"How do you wind up divvying up spaces with other trappers?" I ask. "You said earlier that this is all public land, and anyone can do

Continued on next page

what they want, but people won't come where they know that you are trapping out of respect for you. But how does that work? How does a person get that respect? How do you know who's trapping?"

"If I see someone trapping there, then I won't trap," Bassett replies. "That's basically it. Now, over the years, I've seen where I've set traps, and guys would set traps six inches from me." They figure that Bassett knows what he is doing.

"That's not really that uncommon to have something like that happen," Bassett continues. "Like I told you, since fur prices are down, it basically weeds out the riffraff. You get guys who trap and like to do it, and they respect one another. But as soon as the dollar gets involved, then you see the change."

Trapping is both art and natural science. The good trapper is a student of the environment and animal behavior. The successful trapper knows where, when, and why an animal will follow a particular route, whether it's along a river's edge or on top of a rock outcropping.

"When you set traps, you seem to look for places where geography funnels things," I say.

"Absolutely," Bassett replies. "They're kind of like people. We follow the easy path, as opposed to going up over a hard path, and animals do the same thing. Every animal just takes a little different route."

"When you say you 'know where they are going to go,' the muskrats and the beaver, is that because you've spent a lot of time prepping beforehand, studying them, studying the habitat? Do they have game trails?" I ask.

"You can look at a place in the water," Bassett says, "and I can tell you exactly where that beaver is going to swim. Or a muskrat. Those are two animals that follow habits all the time. If you know what you're looking for, you can just look and see that a rat will swim here or a beaver will. 'Coons are similar that way, but they are a little more diverse. And bobcats—they go exactly the same trail, but they take a long time going there. They might take a month at a time, then, to make a circle. But a bobcat will almost always come back to where you've cut its track before."

Bassett has a variety of tools in his trapping bag. Besides the various kinds of traps, each designed for a different animal or habitat, he uses different scents to lure mammals, buckets for baiting, and stakes with heavy-gauge wire attached to them and the traps to make sure that, if an animal springs one of them, he can retrieve it. Each animal, too, requires a different approach. Some like visual attraction because it catches their attention, while others steer clear of obvious sets and can be brought in only by the scent of meat, whether fish or another mammal.

What goes on in Bassett's mind when he sets a trap? "Well, my goal is to get the animal to come to that trap, step in the trap with either the front foot or hind foot, depending upon the animal," Bassett explains. "And then, if you want to catch it on the front foot, and then if you want to...make a drowning set, do it with all the terrain where you're at so that you can set it appropriately. Either drown the animal or have it pull back away out of the open site so that people and other animals can't see it. And you have to look at that each time you set a trap. Anyway that's what I do. I don't just [haphazardly] set one. It might look like I do, but every time I look to see that the muskrat can drown, or I use a particular type of trap that it can't drown so that it won't twist off. Same with raccoons."

"What about trying to make it look natural? That's part of the process, too?" I ask. "You're trying to disguise the trap at the same time, right?"

"Yeah, you disguise the trap, but that's different," Bassett replies. "On some animals—they hunt strictly on eye appeal. Bobcats are one. They go for eye appeal. A raccoon has a real good nose. But they look for eye appeal. So, if you can get something there that attracts

Two adult muskrats add to his overnight catch. While fur prices vary substantially, Bassett will get from $2.50 to $3.00 for each pelt.

them, either a hole or a nice-looking bait, that will attract a raccoon right in."

"But with the canine," Bassett continues, "a coyote, for example—if you have anything flashy like that, unless it's a real dumb one or a pup, they'll usually shy right away from it. Every animal is different. A lot of them like flash and glitter, and others absolutely none."

Regulations stipulate how often a trapper must check his traps, partly in response to public agitation a few decades ago about animals that were left to suffer slow deaths in traps that hadn't been checked in weeks. Utah state regulations require that a trapper check leghold traps every forty-eight hours and a killer set or drowning set every ninety-six hours.

"You need to check your traps," Bassett says. "But if you're trapping anyway, you want to do that so that if you have something, there's a good chance something will come by and eat it if you don't take care of it. So you want to take care of it that way, and just for the animal so that he doesn't suffer unnecessarily. I think that's one of the worst things that I can see is people who set up traps and then don't check them. It's a good law, although trappers really

Continued on next page

should police themselves, and most do, but you know, with everything, there's a few bad apples in everything you're involved with."

Actually Bassett checks his traps more frequently than the law requires. One of the reasons he's a roofing contractor is because it gives him the time and freedom to set his traps one day and check them the next. He also harvests every part of an animal: its pelt, its scent glands, and its meat if it is appropriate to use as bait to lure a predator.

Bassett and I talk about the touchy subject of knowing how many animals to take out of an area. After all, the fur trappers of the 1800s didn't practice game management, often taking every animal they could because they didn't expect to return. Bassett says that if he doesn't catch anything for a while, he'll pick up the traps, or he'll pull out if he thinks he has caught enough.

How does Bassett know when he's taken enough? What are the signs? "If you don't see any more activity," he says. "And then it depends upon the population."

"In that area we looked at yesterday where we saw so many of those 'beaver thoroughfares,' how many do you think there would be in that area?" I ask.

"Working through there," Bassett replies, "there would be, I'm guessing, eight. Eight to ten. And that's from the dam up to almost that last set," an approximately two-mile stretch.

"So, out of that eight in that area, that two-mile stretch, what would you feel comfortable taking out of there before you say, 'That's enough'?" I ask.

"Four or five," Bassett responds quickly. "And then you can monitor it. It depends on the size of them also."

Does the gender of the animal make a difference? "Well, if you catch an overabundance of females…," Bassett's voice momentarily stops as he senses the problem. "But that's usually after the fact with beaver [they are killed by the traps]. If everything I caught was females, I might want to pull out a little quicker."

"So, if you had three females at that point, you might decide that's enough?" I ask.

"That might be enough, yeah. But you kind of watch the environment and see," Bassett explains. "Now I don't trap up above here farther [his trap-line terminus in Blacksmith Fork Canyon]. And these beaver will feed into this area every year. In fact, I could probably trap everything out of here, and next year there'd be the same amount of beaver because the ones from above funnel down into it."

"But I don't do that," Bassett continues. "And then I'll go and trap this little section here, and there might be really good sets up here a block—and I know there are, really good sets—and I'll never touch them, just to let them rest."

This kind of knowledge only develops over time. Less-experienced trappers than Bassett aren't as attuned to the dynamics.

"And that's a problem that you have," Bassett says. "If I go into an area like here, and I take out the allotted beaver, and someone else comes in, and the beaver will still make signs so it looks like there are still quite a few beaver, and then they'll take two or three. And then maybe somebody else will take two or three. And then, before you know it, your population is gone."

Bassett has seen that happen before.

"Whereabouts?" I ask.

"Everywhere I've ever been," Bassett replies, his enthusiastic voice trailing off for the first time in our conversation.

Field Results

The Friday and Saturday of the first weekend in November 2002 when Bassett and I trapped are unseasonably cold. Snow had fallen the night before, and the temperature had dipped to twelve degrees. While the snow helps him see tracks and gauge movement, it also means traps have frozen in and animals would be sluggish. Bassett reflected on what it must have been like for the earlier fur trappers in Cache County.

"It would have been brutally hard," Bassett observes. "First off, we have access. If we want to go up Logan Canyon and go trap some beaver, we can hop in our truck or car and drive on up. They had to go by horse. And, when I go into that pond, I can put on my hip boots or waders, and they didn't have that. Now, if you wade into it today—this morning—it'll be cold, but you will warm up. The trails and that, the brush—if you go into any wilderness area where they don't have a road, and especially if you pack out a beaver, a sixty- or seventy-pound beaver, you know what they went through."

Bassett plans to set thirty traps in Blacksmith Fork Canyon and about twenty in Logan Canyon; the streams in both canyons are tributaries of the Bear River. The first traps we set are near the Bassett family cabin on Blacksmith Fork. Bassett sets traps for beaver with a submerged Conibear; for raccoons, with a Conibear in a covey trap camouflaged inside a rock den and baited with brightly colored apples, squash, and scent smeared on a rock, with a stick over the top of the opening to force the raccoon down into it; and for raccoons and minks, with a leg-hold pocket set two inches under the water and placed in a small cove that Bassett had previously dug out of the riverbank. The pocket set also is baited. The minks won't be attracted to the fruit but to the lure and the pocket itself due to their innate curiosity. The lure is a "sweet" one for animals like raccoons, minks, and beaver.

Farther upstream, Bassett had agreed to set traps in Hyrum City Park to get rid of some raccoons that had been getting into garbage cans. He has anchored a plastic bucket in a rock enclosure in a deep thicket of willows. He has been baiting it for several weeks before he set his first Conibear trap. Bassett also has set a snare in the park, a cable with a system that locks down on the animal's neck as he goes through it. The snare is set for a raccoon or fox, although a bobcat also could be caught in it.

After looking at the depleted cattails and rushes due to drainage work on the river, Bassett decides to put out fewer traps for muskrats than he had anticipated. With less water in the river, there probably will be fewer muskrats since they have become more vulnerable to predators. He sets his traps for muskrats where he can see *feed beds,* areas where cattails have been chewed. These feed beds are crevices in the riverbanks where the muskrats naturally come up into the cattails and begin to feed in their "kitchens," as Bassett calls them. He is concerned about setting the muskrat traps so that he won't get a beaver. The soil was too soft, and he wouldn't be able to stake the trap solidly enough to hold a beaver. Bassett uses a stop-loss trap and a one-and-a-half with a weight for the muskrat in the cattails.

At the next stop, Bassett sets out a snare for red foxes. But he notes that, while in the past he could catch twenty or so foxes in the area, even when not trapping for them too hard, there are fewer now. Coyotes, which are expanding their habitat up the canyon, have driven them out. They compete with foxes for the same prey but more efficiently. Also the coyotes go after the foxes.

The next day, we retrace our steps, stopping at the family cabin. Of the three traps, Bassett has picked up one beaver, a kit. He is noticeably disappointed that it is a young one, but he also feels it is one that had been doing a lot of damage.

The second stop has a different set of stories to tell. In the muskrat set, Bassett has caught a 'rat, but only the partially eaten remains are left in the trap. A raccoon has come by and taken the muskrat into a thicket for a meal, compliments of Bassett. He resets the trap to catch the raccoon, guessing that

Continued on next page

it will remember its success the night before and come back to the set.

The third place also shows activity. In the bucket buster, the Conibear has sprung but has only grabbed a few swatches of raccoon hair. The animal had gotten away. At the pocket set, where he had placed a leg-hold trap and been looking for either a muskrat or a beaver, he has caught another kit beaver.

"So it's just like another aspect of nature. It's the younger ones that get caught first by the predators, right?" I ask.

"Yeah, pretty much so," Bassett says. "They're inexperienced, and a big beaver on that set there wouldn't have gotten caught because he's too wide for that pocket. Now this little one, he just fits in perfect."

At the next stop, two out of the three traps have muskrats in them. All have drowned. The mature muskrats are about twelve inches long, pretty good size, according to Bassett. The other trap has been frozen in and rendered inoperable.

Bassett had set five traps at the next location. In one of them is a shore bird that unfortunately has stumbled across the trap at the edge of the cattails. Two of the traps have caught muskrats. The Conibear for the beaver has been sprung, probably by a muskrat swimming through, Bassett figures.

The next stop has no animals, nor have any traps been sprung in spite of the fact that the area had a lot of beaver sign on the banks, especially where they had dug trenches back into the thickets of reeds.

When we return from checking the traps in Logan Canyon, where we found nothing, Bassett reflects on the average success of these two days. "Stuff happens like that raccoon getting that rat today. That's just part of the trapline experience, though. You have to accept certain losses," he says.

"Your loss was the 'coon's gain, at least that day," I say.

Then Bassett adds quickly, with emphasis and a short laugh, "We'll get the 'coon. We'll see what tomorrow brings."

Transportation and Anglo Emigration

One foot followed another in the historical paths of the Intermountain West. First, buffalo grazed the emerging grass; then Indians made trails on top of those patterns as they stalked the ungulates. Later, fur trappers stepped into the prints of moccasins. These routes became the roads of migration into the West.

The Oregon Trail, which the Indians called "Great Medicine Road," reversed the earlier movements by Jedediah Smith up the Bear River from Soda Point to the Uinta Mountains. The first emigrants signed on for passage with the fur caravans, and trappers became the first guides for their wagon trains. Dr. Marcus Whitman's trip in 1836 to found a mission at Walla Walla is generally credited as opening the wagon path along the Bear River that eventually became the Oregon Trail. Other Protestant missionaries followed and brought cattle with them. Today U.S. 30 practically overlays the Oregon Trail.

John C. Fremont entered the floodplain of the Bear River near Evanston, Wyoming, on August 21, 1845, and followed the Oregon

The Oregon Trail paralleled the Bear River through Bear Lake Valley in Idaho.

Trail north on his way to Great Salt Lake. By that time, migration along the trail was reaching its peak. He waxed rhapsodic in his report to Congress over the sight of wagon trains encamped along the Bear River, perhaps because of the loneliness and wistfulness he had been experiencing in his explorations.

> Crossing, in the afternoon, the point of a narrow spur, we descended into a beautiful bottom, formed by a lateral valley, which presented a picture of home beauty that went directly to our hearts. The edge of the wood for several miles along the river, was dotted with the white covers of emigrant wagons, collected in groups at different camps, where the smokes were rising lazily from the fires, around which the women were occupied in preparing the evening meal, and the children playing in the grass; and herds of cattle, grazing about in the bottom, had an air of quiet security, and civilized comfort, that made a rare sight for the traveler in such a remote wilderness.

Life on the Oregon Trail was difficult and often sad. Small burial sites like this one at Soda Point, dutifully fenced if only to contain a fading memory, speckle the margins of westward migration routes.

There was the threat of Indian attacks as Anglos challenged the sovereignty of Native Americans and began laying legal claim to land. Yet things were relatively peaceful in 1845 along the Bear because the Blackfeet had retreated to the north. Fremont recorded one incident, though, that revealed the ongoing tension between Anglos and Indians, when a party of Shoshones in full war regalia galloped toward his group.

Apparently they had been aroused by Fremont's American flag, which the Sioux and neighboring Indians often displayed during their attacks on the Shoshones. Fremont was mistaken for an enemy, but the chief calmed the warriors, and Fremont went into the camp to trade and sample the odiferous *kooyah,* or tobacco root.

One of the few pleasant distractions for emigrants along the Oregon Trail was what the fur trappers called Beer Springs near the place where the Bear River begins to turn south around the Bear River Range. Fremont remarked how he looked forward to seeing the site that emigrants had been talking about. He described the springs blowing out of spouts to a height of approximately three feet like a *jet d'eau.* "It is a hot spring, and the water has a pungent and disagreeable metallic taste, leaving a burning effect on the tongue," he wrote. Fremont also noted a hole nearby with hot gas escaping at regular intervals. When he tried to get an emigrant boy to smell the gas, too, to confirm his perception that it was nauseous, the boy refused "with a few indistinct remarks about the devil." Fremont then analyzed the water and found it composed mostly of sulphate of magnesia with portions of sulphate of lime, carbonate of lime, and carbonate of magnesia.

Of course, the emigrants had never seen Yellowstone so they could be excused for their misplaced exuberance for the soda springs. Don Mabey quotes the naturalist John Kirk Townsend, who was traveling with a missionary party, foreseeing a time when the area would become a haven for thousands "from the gay and fashionable world, as well as Invalids and spectators," who would be drawn to the effervescent waters. Washington Irving, in his recounting of Bonneville's exploits, said people drinking the waters indulged "in a mock carouse; quaffing, pledging, toasting, bandying jokes, singing drinking songs, and uttering peals of laughter, until it seemed as if their imaginations had given potency to the beverage, and cheated them into a fit of intoxication." Emigrants reported making successful biscuits. Also in the area of the springs were patches of white clay that the Indians used for soap or to cover themselves, "whitening skins," as the emigrants described it.

Today Beer Springs bubbles without fanfare underneath Alexander Reservoir, one more victim of water reclamation projects in the West. Only its namesake, the town of Soda Springs, whose name was altered to appease teetotalers, remembers.

By the 1850s, explorations of the West revealed there was a wealth of other natural resources besides fur. Fashion had changed, too, with the beaver hat falling out of favor in the salons of Europe and the cities along the East Coast. Fur trappers gave way to new entrepreneurs. Men discovered gold in them thar hills, along with silver, lead, and copper. Grazing seemed like another way to make a buck off vast tracts of western land, and a few stout souls thought there must be spots close to rivers that could be tilled with water drawn from them. But first there needed to be ways to move the ore, cattle, and grain across those rivers.

Empey's Ferry, on the Boise Ford of Bear River near the current town of Deweyville, Utah, in Bear River Valley, opened for business in 1850. Like other ferries, it was used during the spring and early summer during high water flows and, for the most part, served provisioners for mines in Montana and emigrants on the California Trail.

William Empey, who had received a grant from the state of Deseret, operated the ferry. In what would no doubt be decried in the twenty-first century as improper government involvement in the market, the territorial legislature stipulated that Empey and his partners could charge from three to six dollars per vehicle, depending upon weight, with pack animals fifty cents each, all other animals twenty-five cents, and pedestrians ten cents. Empey later sold the ferry to Abraham Hunsaker but remained active in the ferry business when he teamed with three others to receive another territorial grant in 1852.

A streambed of cobbles made Hampton Ford a desirable place to cross the river, either for east-west traffic to California or north-south traffic to the later gold mines in Montana.

Approximately six miles upstream from Empey's Ferry, Benjamin Hampton and William Godbe opened another ferry in 1853 near what is now the town of Collinston. Earlier, Indians and fur trappers had used the ford because of its gravel bottom. Wild hay grew along the river there, so it was a natural resting spot for wagon trains that needed to feed their horses and livestock. The Hampton Ferry and Ford accommodated both east and west California traffic during the Gold Rush years of 1849–52 and the north and south traffic that came later once gold was discovered in Montana. It became the primary ferry, and when the Stage Coach Inn, a rock hotel, was built at the site, it also became a stopover for three different stage companies. The ferry service was sporadic after 1859 and ended completely when Hampton and Godbe built a bridge in 1866 and began charging tolls.

Sometimes entrepreneurs constructed ferries at natural fords. At other times, wagon trains sought out other fords where they wouldn't have to pay fees. It became a cat-and-mouse game, with wagon masters trying to stay one step ahead of the ferry entrepreneurs.

Fred Selman, a Deweyville rancher and Soil and Conservation Service board member, responded to a call from Governor Scott Matheson requesting contributions to the events surrounding the Utah centennial celebration in 1996. Relying on an 1849 record that he and his wife, Laura, had found from the Stansbury Expedition to survey the Great Salt Lake for the federal government, the Selmans set out to locate a Bear River crossing reportedly "seven miles, 184 feet north of a hot and cold springs," now Crystal Springs resort in Box Elder County. Driving north in their car, they found old wagon tracks on a hillside on the Sorenson farm at a spot within thirty feet of Stansbury's survey entry. Moreover, Sorenson showed them an oxen chain he had found on the riverbank that would have been used to hook up an additional team to cross the river or haul a wagon up a steep slope. It convinced them that the so-called Bear River Crossing was a third ford, in addition to the Boise and Hampton ones. The Bear River Crossing picked up a lot of emigrant traffic, too, as a departure point for the Salt Lake Cutoff on the California Trail.

Ferry traffic up and down the Bear River, from Montpelier, Idaho, to Bear River City, Utah, was short lived, though, because of what everyone knew was coming down the pike—the railroad. It was a heady time, and land developers and merchants built towns almost overnight in anticipation of wealth that would follow the locomotives. The town of Corinne, Utah, was one of them. It seemed to be in the right place at the right time.

The town incorporated in 1870 shortly after the driving of the golden spike. There was a water system in place by 1869 with a steam-generated pump that drew water from the Bear River and stored it in a wooden settling tank. Water then flowed throughout the city via a wooden pipeline. Unlike other towns like Salt Lake

The view for emigrants looking west at the Boise Ford, one of three fords on the Bear River.

A rock sign marks the Boise Ford and the embarkation point for Empey's Ferry.

City that were platted along cardinal points, Corinne was laid out parallel to the railroad tracks.

There was another major difference between Corinne and Salt Lake City. By the 1860s, increasing numbers of emigrants who were not Mormon were settling in the Intermountain West. These *gentiles* were concerned about the growing, powerful theocracy whose epicenter was in Great Salt Lake City. They feared that its communal organization and brand of economic socialism might thwart their capitalistic dreams. So they saw Corinne as a potential new center that would compete with Salt Lake City for the title of crossroads of the West. Corinne would be a gentile town and become the trading center for the Great Basin. There was ample reason to support that grand vision, too, because the Union Pacific railroad was considering three towns for its junction city: Ogden, Bonneville, and Corinne.

In his *The History of Box Elder County*, Frederick Huchel notes that in 1869 Corinne immediately began promoting itself and convinced the Union Pacific to survey the town for plats, with the railroad getting every other plat in return for service. Corinne felt it had the edge because of the Bear River, which at the time was thought to be navigable. Unfortunately, from 1862 to 1872, Great Salt Lake was rising. So the river soundings indicating the Bear River was eighteen feet deep and three hundred feet wide at Corinne were misleading. Further encouraged by the knowledge that the town was only six miles as the seagull flew from the lake, entrepreneurs were convinced that Corinne, nicknamed the "burg on the Bear," would one day become the "Chicago of the West."

They were sure that a town where the railroad crossed the Bear River was a logical choice for a hub. At the time, Corinne was near the northernmost point on the railroad. It was naturally situated to be the center of trade with mines opening to the north in Montana and Idaho. Ore could be brought by wagon to the Bear River, where it could be loaded on the railroad and sent to smelters. Other grand ideas followed. Why not begin steamship trade on Great Salt Lake and bring ores from the Oquirrh Mountains approximately sixty miles to the south to the railroad via steamship? Farmers in the area could use the railroad, too, to move their crops grown on the fertile lands in the floodplain of the Bear River Valley.

Early trade seemed to bear out the braggadocio. About eighty wagons with four hundred mules arrived in Corinne daily; the six-hundred-mile trip from Montana took ten days. They crossed on the Bear River upstream at the ferry sites at the Boise and Hampton fords. Ore came south; food and supplies went north with prices like two dollars for a dozen eggs and a dollar per pound for sugar. Farmers came from Cache, Marsh, Gem, and Bear River Valleys with wagons of grain. The traffic was so heavy that wagons sometimes had to wait for days and weeks to unload.

Colonel Patrick E. Connor, who challenged Brigham Young from his command post at Camp Douglas in the foothills overlooking

Salt Lake City, built a smelter on the banks of the Bear River at Corinne for the ore coming from Montana. Later, he decided the same smelter could be used for ore from the Oquirrh Mountains. But he needed a way to get to the southern shore of Great Salt Lake. He needed a steamship.

At first, the *Kate Connor* only carried passengers, charging them five dollars for a round trip between Lake Point, on the southern shore of the lake, and Corinne. The steamship's route went down the Bear River to Bear River Bay and then along a channel between Fremont and Antelope Islands to Lake Point. But since the river meandered through its delta after leaving Corinne before dumping into Bear River Bay, it actually took thirty-five miles and several hours to reach the lake.

Connor eventually ran three steamships and schooners, hauling telegraph poles and railroad ties for the Central Pacific besides silver ore. But the citizens of Corinne, filled with public pride or worried about an army officer controlling their future, decided they wanted their own steamship. They began raising funds for the *City of Corinne,* a three-decker steamship capable of carrying three hundred tons, larger than any of Connor's ships. It was built on the Bear River at Corinne, with its boilers and engines sent around Cape Horn to San Francisco, and had riverside docks as part of the infrastructure. At the first attempt to launch it in March 1871, the ways the ship was supposed to slide down into the river sank into the mud instead. It took seven more hours to free the rear-wheeler.

The *City of Corinne* took lumber and supplies to mines opening up in the Oquirrhs and wire for the Western Union telegraph at the south end of Great Salt Lake, and brought back ore. Initially it made three trips a week, with arrival times in Corinne scheduled to meet the Central Pacific's westbound passenger train.

The lumber came from upstream on the Bear, largely trees cut around tie-hack camps in the Uinta Mountains and intended for railroad ties. In their "The Navigational History of the Bear River," C. Gregory Crampton and Steven Madsen note that mine props, stope stakes, railroad ties, cordwood for charcoal for smelters, and other lumber products were floated down the Bear River to support the railroad and mining industries. Communities sprang up along the river from 1867 to 1935 to support these activities.

Loggers cut the trees and floated them down the river to Evanston and Bear River City, Wyoming, with the most business occurring before the driving of the golden spike and shortly thereafter, throughout the railroad-building era of the 1870s. The trees, usually lodgepole pine or Engelmann spruce, were hand-cut in Union Pacific sections in the Uintas. They were skidded or hauled to the river in winter on wagons or sleighs. Then, when the ice in the river melted, logs floated downstream with the spring and early summer runoff.

Sometimes, though, the river didn't cooperate. Alexander Toponce, a well-known freighter, contracted with the Central Pacific Railroad to haul a hundred thousand railroad ties down to the Bear

(above) Both the Union Pacific and Central Pacific Railroads relied on tie-hack camps like this one on the East Fork drainage in the Uinta Mountains to supply them with lumber.

(below) Remnants of a flume used to float logs down the Bear River during swollen spring runoff.

River between Montpelier and Soda Springs and float them down to Corinne. The contract stipulated that Toponce deliver the ties before either the Central Pacific or Union Pacific crossed the Bear River. Toponce used fifty oxteams to haul the ties down to the river. But the Bear froze up earlier than usual that winter and locked the ties in the ice at various places for a hundred miles. Toponce couldn't get them to Corinne on time. When he finally muscled them down the river, the Central Pacific refused delivery, and Toponce had to stack them on the banks of the Bear River at Corinne, where they stayed until they were used to build the Utah Northern Railway.

It was that Utah Northern Railway, completed in 1870, that rang the death knell for the dreams of Corinne. There never was enough freight-hauling traffic on Great Salt Lake to support Connor's ships and the *City of Corinne*. By July 1871, trips by the *City of Corinne* had dropped to twice a week; in August, the steamship went on an as-contracted basis. In 1873 the *City of Corinne* had to moor

The Methodist-Episcopal Church in Corinne dates from 1870 and is the oldest Protestant church building in Utah.

in Great Salt Lake because the ship couldn't get up the Bear River; sandbars were forming at its mouth due to the rising lake waters. The owners tried to be flexible, basing the ship at various other spots on the lake, starting with Promontory Point and then Kaysville.

When the lake trade petered out completely, the burghers of Corinne went back to focusing on the railroad. But events began chipping away at that business, too. The wily Brigham Young had been ahead of their game. When he saw the inevitability of the railroad entering his kingdom and its commercial advantages, he built a railroad from Salt Lake City north to Ogden, and that city became the junction and transfer point for railroad traffic. Young had trumped the gentiles of Corinne. Finally, the wagon trade dried up completely when the Utah Northern line was extended north from Ogden into Idaho east of Corinne, and the wagon-freighting business became another relic of the Old West.

The building at 3975 and its neighbor were built in 1864, the beginning of the heyday of Corinne. According to Diane Clifton, a 30-year resident of Corinne, 3975 once housed a bar and liquor store and later a bookstore. The adjacent building was the post office and barbershop. There was a brothel behind the buildings, as well as one directly across the street. Clifton says her investigation of Box Elder County records shows that the buildings have been sold for as little as ten dollars or as much as ten thousand, depending upon real-estate speculation at the time. City lore has it that the buildings have been won and lost in card games seven times over the past century and a half.

Anglo Settlement

The Homestead Act of 1862 opened up Bear River valleys to settlement. Any adult or head of a family who had not borne arms against the government could apply for 160 acres of land in the West. If the person lived on the land, worked it for five years. and paid a registration fee, the land would be his.

Typically the first settlers lived on small creeks on the benches rather than on the Bear itself due to the difficulty of lifting water from the river. Gravity could move water from the creeks more efficiently. But during periods of drought, settlers had to haul water from the Bear River with a horse in barrels using a *go-devil,* a device consisting of two boards under logs functioning as runners with a barrel on top.

The first settlers in that broad plain now called Gem Valley were drawn to its fertility, especially in the southern sections where the rolling land formed mounds less filled with volcanic rock, just as the mountain men like Ogden had found it attractive for trapping some fifty years earlier. But it wasn't always known by that name. Originally it was called Gentile Valley, a name sometimes used for its southern extension near Thatcher. Some say it was called "Gentile Valley" after the gold miners in the Caribou Mountains north of Soda Springs, who wintered in the milder, southern portion of the valley. Others say the moniker had more to do with religious tensions at the time.

Robert Fellows, a gentile or non-Mormon, made the first settlement in the valley in 1865 on Williams Creek. According to the [LDS] *Church News,* bitter "anti-Mormons" began settling in the valley and swore to keep out Mormons by not selling land to them. It was a weak alliance because some reneged, especially when Brigham Young began sending missionary families to colonize the valley in the 1870s. But tensions ran high, and some Mormons were driven out of the west side of the valley to the east by drawn gunpoint. The west side, then, became the gentile side, and apostate Mormons congregated there. The Bear River, which meanders down the center of the valley, became the Mason-Dixon Line.

When the Mormon postmaster moved away, Harry King, a Mormon antagonist, became postmaster and moved the post office from the original Cottonwood Creek/Bear River confluence to a site farther upstream. He renamed it the "gentile post office" to broadcast the gentiles' desire to control that side of the valley. According to Vivian Simmons and Ruth Varley in *Gems of Our Valley,* King would not deliver mail if it was addressed to Mound Valley, which was the Mormon enclave. Later, settlers vowed that no Mormon meeting house could be constructed on the west side. But Solomon Hale, a Mormon, opened up his house when he heard that pronouncement. As years went by, non-Mormons began selling their land to Mormons, and a Mormon ward was set up on the west side of the valley. By 1910 peace reigned, and residents officially named the area Gem Valley.

Downstream from Gem Valley, settlers were drawn to Box Elder County and Bear River Valley by the availability of irrigation water through the Bear River Canal Company, which owned some of the most senior water rights on the river. However, brochures touting the area suggested that intensively irrigating five acres would be more profitable than trying to dry-farm 160 acres as settlers were struggling to do in the rockier, more-arid northern Gem Valley, where the hay harvest was meager.

An 1899 promotion sponsored by the Oregon Short Line, Union Pacific, and Northwestern Pacific railroad lines encouraged people to emigrate to land owned by the newly incorporated Bear River Land, Orchard and Beet Sugar Company. Land from the company was forty dollars an acre, including water rights from the Bear River Canal. Water rental was a dollar per acre per year "except orchards over five acres, and on them the annual water rental will be $2.50 an acre after they are 5 years old." The minimum purchase was five acres, "one quarter in cash, the balance in two, three and four years, interest at 6 per cent, per annum, payable annually."

To that point, the brochure was up front. But when it began promoting the climate and other benefits of settlement, it probably would have run afoul of truth-in-advertising laws if they had existed then. Of course, there were none in that period of caveat emptor. Here's what the brochure boasted:

> Utah's climate is all that is claimed for it by the medical fraternity of the United States. The action of the climate in the Bear River Valley is that of a strong stimulant tonic to the whole constitution, and particularly to the nervous system. It is enjoyed by those in good health and adds a zest of life that is never found in lower altitudes, and those suffering from pulmonary troubles are greatly benefited, owing to the dryness of the atmosphere; cool nights; plenty of sunshine.

It probably could have stopped there and not been too off base. But then it started to sell snake oil:

> It is difficult for one who has never experienced the effects produced by coming from a low to a very high altitude, to understand the wonderful exhilaration that follows breathing this dry, rarefied air. Those in health feel and enjoy the powerful tonic, almost intoxicating effect, but not to the marked degree that the invalid does. It seems to supply a necessity in his case. Nor are the effects limited to a temporary stimulation, to be followed by a corresponding depression. Very soon his appetite is better, his digestion improves, the feeling of lassitude disappears, exercise becomes a pleasure and he is able to enjoy the outdoor life to which he is invited by new scenes, sunny skies and bracing air.

Bridges on the Bear River are colored in controversy stemming from the nasty cultural wars of the middle 1800s. The original bridge between Corinne and Bear River City was torn out because Mormons and Gentiles didn't want to have contact, according to Charles Holmgren, whose emigrant convert great-grandfather was a blacksmith in 1869 "in the den of iniquity" that Mormons labeled Corinne. The current, rickety metal bridge at the east end of Main Street was dubbed "the hanging bridge," according to the salty Diane Clifton. "They used to hang a Chinaman every night just for sport. And any Mormons who came over from Brigham City and preached got hung, too."

> With the increased exercise there is a more rapid gain in weight and strength. These are the usual effects in appropriate cases, the improvement being modified by as well as modifying the local part of the trouble.

Later, the brochure made an attempt to add a disclaimer, assuming the reader would know what it meant. It identified people who should not come to the Bear River Valley as those who "already do not have healthy lung tissue and who already have 'degenerative heart changes.'" It also said folks who had problems with "pleuric adhesions, enchymatous degeneration of the heart and lesions of the vital organs" might want to think twice about emigrating.

The pièce de resistance in local salesmanship came when the brochure talked about "natural homes" in a place that was the apple of God's eye and where business was next to godliness:

> Coupled with the climatic advantages offered the health-seeker in Utah, there are excellent opportunities offered to home-seekers and investors, especially those who want a place where the business conditions are not only favorable, but where they can enjoy, while at work, pure, wholesome air and where, when [it] is done, they can have the benefit of all of Nature's benefactions. The former will promote business, success, and the latter will complete the measure of happiness and prolong life. Think of having a home in such a place! A home in which you can gradually drift into green old age; a home filled with the voices of children and children's children; within which you can retire at evening, watch the fanciful figures that come and go in the blazes of the grate, enjoy a feeling of great security, and be at perfect peace. If you would have all these beatitudes, select Utah for your home. There the Almighty has been prodigal with His gifts.

On second thought, you can read copy not too different from this in the weekly real-estate sections of Intermountain newspapers. The only difference? Real-estate ads today never talk about where the water for subdivisions will come from. A source of water for households is assumed, and future hookups are cheerily assured.

The primitive encampment at the May 2003 Cache Valley rendezvous on the Left Hand Fork of Blacksmith Fork, a tributary of the Bear River. The rendezvous carries on the tradition of the marketing plan successfully implemented by William Ashley and the American Fur Company in the 1820s and 1830s.

Cache Valley Rendezvous

Every Memorial Day weekend, a loosely organized group of throwbacks retreating from the present meets to remember the mountain-man rendezvous. Some make an avocation of traveling around the country to various celebrations, while others refuse to stray from the valleys where it began. In May 2003, they gathered for the twenty-seventh time to reenact the Cache Valley rendezvous on the Left Hand Fork of Blacksmith Fork, a tributary of the Bear River.

Trying to retain authenticity isn't easy, and there was visible tension in the encampment as the itinerants adapted to the demands of the twenty-first century. On the right or southeast side of the stream was the "primitive area," a field dotted with white clusters of tipis and tents. On that side of the creek, everything was supposed to be the way it would have been in the 1820s and 1830s. The idea was to reenact the activities of the original William Ashley group.

On the left side of the creek was what some derisively called "the tin tipis." These were the motor homes and trailers, and while some of these folks spent most of their time in their encampment, not mixing much with the primitive group, there was some intermingling. While some of the primitive campers were sarcastic, others nodded toward the site with goodwill, saying, "They're simply older mountain-man people, and they've paid their dues. They have the mountain-man spirit, but they just can't live like that anymore."

The Cache Valley rendezvous is organized by the Old Ephraim Mountain Men (OEMM). They try to keep rules to a minimum, and most are there only to make things as authentic as

Continued on next page

Barter is the preferred method of doing business at the rendezvous. Fred "Medicine Hands" Nudd takes a look at the dark amber offered by Alan Riser.

Merchants offer a variety of herbs and dried foods, along with primitive tools, on Traders' Row.

possible. One rule is that in the primitive area, you only are allowed to bring in your car on the first day, Friday, to get things set up. At a certain time, cars have to be removed from the primitive area, and after that, they can't return until Monday morning, when it's time to break camp. While most people bring coolers for ice and jugs for water, the coolers are supposed to be camouflaged under blankets or serapes. If you bring in a Coleman lantern, you are encouraged to keep it inside your tipi.

Like the rest of America in the twenty-first century, there are even subcultures within the OEMM rendezvous. One of the groups is called the Free Winds Trappers Association. They have their own flag and are opposed to most rules and regulations, believing people should come mostly to have a good time. They

Mad Jack sets up his blacksmithing and custom forge business at the yearly rendezvous.

maintain that rules should only exist between parent and child.

Flags were everywhere in the primitive camp, adorning both tipis and tents. They also were for sale. Inside Alan "Many Bottles" Durtsci's three-sided tent were several flags: contemporary American flags, colonial flags with thirteen stars, a Don't Tread on Me flag, an Australian flag, and a few flags from the Confederacy.

Durtsci wasn't the only person with a nickname at the rendezvous. There were Autumn Wind, Spirit Dancer, and Morning Star. Two Owls was the "booshway," the head honcho of the OEMM. Just Sittin' There was, presumably, taking life easy. In fact, there was a "game page" in the rendezvous souvenir program that challenged participants to match the mountain-men names with given ones. This was a time and place where a person from an otherwise urban or suburban life could adopt a new persona. The nicknames had become trademarks; mountain men imprinted them on their business cards. Mad Jack's business card advertised his skills: coyote hunting, leatherwork, horn work, scrimshaw, knife making, welding, sharpening knives and scissors; fashioning buffalo skulls, hides, and parts; blacksmithing, and custom forge work.

Traders' Row was the commercial nexus of the primitive encampment, supposedly duplicating an area with the same name at the nineteenth-century rendezvous. Almost anything could be traded or purchased. Many

Continued on next page

Alan "Many Bottles" Durtsci specializes in flags and hand-painted gourds and ladles in his store.

people had bone-handled knives that they had carved, woodwork, beads, herbs, baskets, all sorts of furs, and mountain-men clothing. Restaurateurs offered food at two cafés. Root Beer Man sold his signature brew and ice cream sodas at the camp canteen.

Although the canteen proprietor kept his establishment dry, other mountain men stored beer in individual coolers. In fact, it was unseasonably hot that Memorial Day weekend; the temperatures hit the high eighties. The excitement was palpable in the primitive camp when someone spied an ice truck coming up the road, selling bags for three dollars a pop. While some people were walking around in the afternoon in their freshly sewn buckskins, it was obvious that period costumes were stifling that day. Authenticity had its price, and most heavily clothed folks weren't moving at all, once they had put up their tipis, preferring, instead, to retreat to the shade.

Fred "Medicine Hands" Nudd from West Valley City, Utah, and Alan Riser from West Corinne were conducting business. Alan was in Fred's camp to pay him back for some muzzle-loading caps Fred had given him. Alan also had brought amber to trade with Fred. Although both of them were wearing clear amber necklaces, which they call *costume amber,* Alan was giving Fred his choice of what's called *darker amber,* resin from trees approximately one million years old. The resin had flowed over darker material, which gave it the darker color. It's a little bit more valuable,

On an unseasonably hot Memorial Day, there is plenty of time to tickle a mountain man resting in the shade.

too. The amber had come from an area around the Baltic Sea and found its way into Alan's hands via an ancient route.

Although barter was the preferred method of exchange, cash was accepted. "But no sales tax is charged here," one trader said with an edge in his voice. "Sales tax is a bad word around here. You know, back in the 1840s, there was no IRS," he added, lumping all taxes into his bulging bag of gripes.

The primitive camp had a place for a council and fire. The OEMM had arranged activities for the rendezvous—a Dutch oven cookoff, musket shooting, knife and tomahawk throwing, archery and spearing, outfit contests, and a Moonlight Madness Sale—as well as a raffle for a blanket with intricate beadwork. But some mountain men said that they didn't like to get together as a large gathering, preferring to wander in small groups to see friends or stick close to the home fire.

Several people said that they could get quite "rowdy" toward midnight and the wee hours of the morning, honoring that tradition of the original rendezvous. A Cache County Sheriff's Search and Rescue Squad command trailer was parked on the tin-tipi side of the stream, prepared for any breach of the peace or other backwoods mishap. But on Sunday morning, there were LDS Church services in the camp, certainly not authentic to the 1820s and 1830s but, rather, an adaptation to the more recent history of the Bear River Basin.

At Last Chance, most of the Bear River flows through a five-mile long pipe, diverted by PacifiCorp to drive the turbines at Grace.

Once able to move at will
I stall,

Your decrees
Divide me
By time and pedigree.

Diverted
I bleed.

From dams
I fall,
My glory in power released.

I return to the sun
Still open and free

Till you shroud me in a pipe.

Soda Point Dam at the northern terminus of the Bear River Range is the second hydropower facility on the river; it has the potential of generating 14 megawatts of electricity.

5

Chapter

Stakeholders Lay Claim to the Bear River and Its Water

Water wars are legendary in the Intermountain West, the stakes aggravated by relatively scant rainfall. Surface water is the prize, and conflicts rage over springs, creeks, and rivers, the conduits of Earth's precious lifeblood. The struggle for ownership of the Bear River's water has been no different. But while battles in some parts of the West rose to the level of bloodshed, the conflict for water rights on the Bear has been marked more by compromise—and ironclad decree.

Essentially one interest controls Bear River water: irrigation. Its rights trump all others, including the second dominant player, the power generator Utah Power and Light (now Rocky Mountain Power) and its domestic parent, PacifiCorp. Together they determine how the water will be used, and their partnership is forged in the legacy and intricacies of western water law.

But in the twenty-first century, new players want a seat at the bargaining table. They may be Johnny-come-latelies, but they do have a growing influence that flows from the changing population and economic dynamics of the New West. Agriculture and hydropower aren't as significant now to the economy of the region. The voices of new water users, recreation and conservation, grow louder. They are the voices of the increasingly urban, suburban, and exurban West, and although the historical evolution of water law makes it difficult to hear them, they do have the power of numbers. Together the old and new stakeholders fight for dominion over the Bear River, and as they exercise their claims, a river begins to lose its personality, its being.

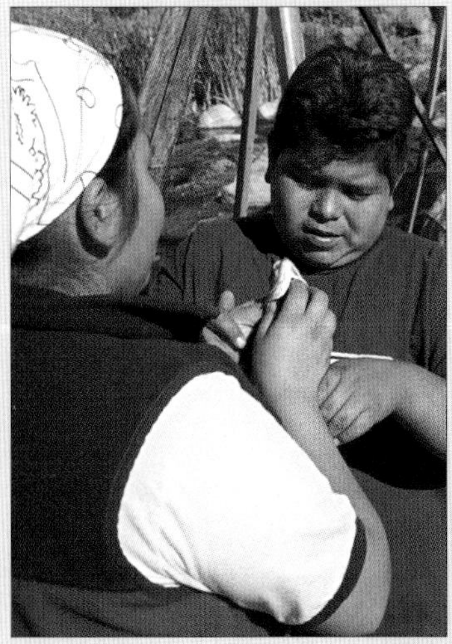

Kids try to unhook a rainbow trout they caught in the spring-fed stretch of the Bear near the Grace power plant.

Harvey Day nets carp at a headgate off the Bear River.

Fishing the Bear

The Bear River never has been known for its fishing, although that could change. There are cutthroat, planted rainbow, and brook trout in its stretches in the Uinta Mountains. A few large, fluvial Bonneville cutthroats ply the semimurky waters below Bear Lake. Trout are found below Soda Point, and there's a put-and-take fishery of rainbows, with some carry-over brown trout, in the spring-fed sections in Black Canyon above Grace power plant.

Fishers would like to gain access to the short, spring-fed stretch between Grace and Cove because of rumors about large rainbow trout. The tailwater area below Oneida Reservoir could become a blue-ribbon trout stream under different conditions, but once tributaries above Preston, Idaho, have dumped their silt into the Bear, the fishing productivity of the river plummets until it gets to the Bear River Migratory Bird Refuge. Then catfishing and taking carp with a bow and arrow become the sports of choice.

Harvey Day from Ogden, Utah, regularly fishes for catfish at the refuge, alone or with his family. I met him when he was netting carp at a headgate off the Bear. He guts the carp, saving only the innards for catfish bait, and throws the carcasses across the road for the foxes and coyotes. He told me that some people don't even bother to discard the gutted fish out of eyesight, leaving them on the ground in public places.

Catfish are king for bait fishers on the refuge. But the wary carp face a new predator: flyfishers, who are attracted to their acrobatics and sheer strength.

But Day and other fishers that spring morning showed no interest in carp as a game fish. In spite of some netted or speared carp going over 10 pounds, they were focused on channel catfish. For Day, a ten-pound catfish would be "medium sized." He has seen ones up to twenty pounds taken from the refuge, although he says quietly that there are bigger fish in the river upstream around Corinne.

Carp never have gotten respect, either from fishers or refuge managers, in spite of the fact that these exotics were transplanted here to be a nutritious foodstuff. They are the bane of refuge managers because their rooting behavior in the canals raises silt that prevents light from reaching underwater plants, which diminishes the biomass for the ecosystem.

Drought is hard on carp, too. Two years into a drought on a dry September day in 2001, when the refuge would be normally low on water anyway, carp in the canals were attempting to swim, flailing forward in about one-foot bursts. Their backs and dorsal fins stuck above the water. At first, I wondered if they were trying to get vitamin D by exposing themselves to the sun, unimpeded by the murkiness of the canal water. But Steven Hicks, assistant manager at the refuge, said that the carp simply couldn't get any lower in the shallow water, leaving two inches of their backs exposed. Because the water was shallow and therefore warmer, it held less oxygen. The carp were struggling to move enough water through their gills to avoid suffocation. The salinity was

Continued on next page

Wendy Burgener goes crawdad fishing in the Evanston, Wyoming, ice ponds, fed by the Bear River, after work. Using uncooked chicken legs as bait, she can catch thirty crawdads in one hour, enough for approximately three-quarters of a cup of meat (she eats only the tails). She throws back the small ones. Pet stores purchase any blue or red ones she catches.

higher, too, due to insufficient water to disperse the minerals.

"They are simply trying to survive," Hicks said, with not too much sympathy, partly because carp are so plentiful in the lower Bear River system. He reported that more than a thousand carp had died that summer in the refuge.

Three twenty-something men in unit two in the refuge were insensitive to the carp's plight. When I drove up to them, they were perched on the dike, laughing as they threw rocks at the vulnerable carp in front of them. The fish couldn't get away because that small pool was all the water left in the unit. One of the men glanced at me and saw my hard look, and his smile drooped slightly as he averted his eyes from mine. They stopped chucking rocks for a moment, until I drove away.

But in 2006, carp are earning new respect. It is startling to be on the banks of the Bear and see and hear a ten-pound fish leap straight up, completely out of the water, and seemingly relish the thunderous splash when its side hits the water. Fly-fishing for carp now is on the upswing as trout fishers, tired of angling for smaller fish or having mastered that aspect of their sport, look for another challenge. Carp are skittish and hard to catch on a fly. But when hooked, they put up a powerful fight. Guides are taking upscale fishers into places where no one ever dreamed there would one day be an attractive fishery. That idea would have drawn a lot of guffaws around the camp table after a day fishing for trout. Now fishers and farmers with private land on the Bear, who can charge for access, providing a new revenue stream to supplement farm receipts, file a new claim on the river.

Bear River Compact

Although the Bear River once flowed into Bear Lake, it began bypassing the lake approximately 11,000 years ago because deposits built up by wave action had created a barrier. Still, waters from the river and lake continued to commingle during flood periods through a natural channel emptying into Mud Lake from the north end of Bear Lake. John Bothwell, an entrepreneur in the late 1880s who was developing a canal system below Bear River Canyon to supply water to the new sugar-beet industry, wondered if he could store river water in Bear Lake to release later in the summer when crops needed it. As reported in Robert Parson's *A History of Rich County,* Bothwell began installing a dam and headgates across the outlet at the north end of Bear Lake. By doing so, he was able to raise the lake level by ten feet.

But Bothwell got into financial trouble, and the idea remained dormant until 1902, when Telluride Power Company looked into it, driven by the vision of Lucien L. Nunn, a lawyer interested in developing water for mining operations. He scouted the area for a hydropower location. In 1906 he came to Grace, Idaho. Because of the unique geography of the Bear as it dropped quickly through Black Canyon, Nunn reasoned that, if he built an upstream dam, he could bring water from it to Grace via a flume and build a head for power generation due to the drop of 530 feet from the dam.

By 1911 Telluride Power Company had excavated the Dingle Canal, bringing Bear River water into Mud Lake. The company installed more headgates at the lake's natural outlet. As a result, it was able to store an additional 25,000 acre-feet in Bear Lake that it released in the fall to generate power at Grace. But its measurements were inexact, and the company actually released more than 40,000 acre-feet.

When Utah Power and Light Company (UP&L) incorporated in 1912 by consolidating Telluride with other predecessor companies, it also took over the Bear Lake project. It installed a pumping system at the north end, now called the Lifton pumps, to control water release more accurately.

Meanwhile, the sugar-beet industry was growing, too, under the mantle of Utah-Idaho Sugar Company (U-I). It labored on a canal system to bring water to Box Elder County and also built a collapsible dam, the Wheelon Dam, in Bear River Canyon to provide power for sugar refining. In December 1912, UP&L and U-I entered into a perpetual agreement where U-I conveyed all its hydropower operations at Wheelon, transmission lines, and pending water rights to UP&L in return for a continuous flow of 900 cubic feet per second of water between May 1 and October 31 and 150 cubic feet per second between November 1 and the following April 30. The water could be used for irrigation or other municipal needs. UP&L also agreed that, if natural flow did not meet the

At Stewart Dam in Idaho, UP&L begins to reroute the Bear River into Bear Lake. The river largely dries up at this point until rejuvenated by an outlet from Bear Lake.

contracted amount, it would store enough water in Bear Lake to fulfill its agreement.

Most of the water rights in that transaction, and in fact most water rights in the Bear River Basin, are territorial ones. The Myers Ranch near Evanston made the first claim on Bear River water, and in Wyoming, in 1862. Bothwell's rights, eventually converted to U-I and the Bear River Canal Company, date from the 1880s. Yet, while these claims seem ancient, after several generations, they are still alive and have more determinative power now than ever.

Major John Wesley Powell was one of the first people to worry about the evolution of water law and development in the West. He was concerned over who should apportion water during droughts and didn't trust the growing power of canal companies. In an 1878 report to Congress, he asked to include a federal law with the earlier homestead acts that would establish priorities and beneficial uses of water. But Congress took no action, deciding to leave water policy to the states and territories.

As the elevation of Bear Lake began to fluctuate, people complained, especially those whose livelihoods, like the resorts, relied on access to the lake. By 1929 the lake had fallen fourteen feet due to water releases for irrigation and hydropower generation. According to Parson, Fish Haven resort was almost wiped out. Residents met to discuss what could be done. They found out they couldn't do much of anything. UP&L had effectively appropriated control of both the lake and the river, and downstream irrigators owned the water stored in the lake. Moreover, hydropower on the Bear River had become critical to the UP&L system. By 1927 UP&L was operating five power plants on the Bear: Soda, Grace, Cove, Oneida, and Cutler (Bear Lake interests argue that Lifton counts as the sixth). Even in 1946, these plants generated 75 percent of the system's capacity.

A federal court ruled in the Dietrich Decree in 1920 that UP&L had the right to divert 5,500 cubic feet per second of water from the Bear River and store it in Bear Lake. Moreover, UP&L was the only entity that could store water in the lake. The ruling confirmed that the power company could store water from a minimum elevation of 5,902 feet, below its intake pipes at Lifton, to a maximum of 5,924 feet, above which waterfront property would be damaged from wave action or wind-generated ice encroachment. That allowed a fluctuation of approximately 22 feet.

Still, stakeholders exerted pressure to control the waters of Bear Lake and the Bear River better. For one thing, people upstream from Bear Lake worried that they would not be able to store water. According to Wallace Jibson, the historian of the Bear River Compact, Wyoming interests were irritated because they felt they should have the first right to store water since Wyoming was the first state through which the Bear flowed after it left the Uinta Mountains. Idaho users felt there was inequitable sharing of natural flow among the three states and vowed that relief should come through the courts or an interstate compact. This friction finally prompted negotiations to resolve quarrels over the Bear River. Congress granted the three states the right to negotiate and enter into a compact on July 24, 1946.

At first, the negotiators attempted to administer the entire river on a priority basis without regard to state lines, but that met resistance. Because the river was already completely appropriated through prior water rights, that would mean there could be no upstream storage above Bear Lake. Two factors became clear. First, there would have to be upstream storage rights that could not take a backseat to those in Bear Lake. Second, the division of natural flow would have to be based on irrigated acreages in the three states, with the priorities established earlier considered but not absolutely controlling.

But storage above Bear Lake was a thorny problem. Where would the water come from? Downstream irrigators didn't want it taken out of their storage water in Bear Lake. The solution was to reduce

UP&L's use of Bear Lake water to generate hydropower. Negotiators would decide upon a lake level below which water could not be released solely for power generation. The level would reflect the average runoff in an earlier five-year period. It would be derived from historical data and good water-forecasting projections.

The trigger level became 5,914.61 feet, 12½ feet above the Dietrich Decree level. Any water below that elevation would be an "irrigation reserve" that could not be used solely to generate power. But UP&L could use any stored water above that level. This would provide 30,000 acre-feet of water that could be stored above Bear Lake, largely in the later Woodruff Narrows and Sulphur Creek Reservoirs. UP&L retained its exclusive right to store water in Bear Lake and could use that water to generate power as it moved down the river to service downstream irrigators. In essence, the compromise fixed UP&L's interests as subordinate to irrigation.

The Bear River Compact was signed in 1958, and a Bear River Commission was appointed to carry out its provisions. The preamble to the compact states that its purpose is "to remove the causes of present and future controversy over the distribution and use of the waters of the Bear River; to provide for efficient use of the water for multiple purposes; to permit additional development of the water resources of Bear River; and to promote interstate comity."

Denise Wheeler, former chair of the Bear River Commission, feels the negotiations were initiated by the three states less because of past water wars and more because of concerns about the way each state would develop the Bear. Ranchers earlier had solved disagreements amongst themselves. The compact carried that cultural tradition of cooperation to the state level.

Besides, the compact really only solidified the power of the entities that already had been controlling the Bear. Once the three states became involved, the players realized that political intrigue could become a factor. It was in each state's best interest to negotiate and allocate to avoid becoming the odd man out. The compact entrenched the status quo, further consolidated western water law, and created formidable obstacles for new claims on Bear River water, ones spawned in the late twentieth century.

The compact became operational during a drought, with annual water supplies ranging downward from 80 to 43 percent of normal, depending upon the drainage. The agreement called for equitable division of natural flow, but in some cases, like Rich County, Utah, that was meaningless because county canals didn't have enough grade to pass on the meager, divertible flow. Wyoming users were still rankled because they were used to little water regulation in their state. Under the compact, they were highly regulated to benefit Idaho, a downstream user, when Wyoming consumers felt their state's watersheds were supplying most of the water. They were not looking closely at the southern perimeter headwaters in Utah's Uinta Mountains. Lawsuits began testing the legal armor of the compact.

The Rainbow Canal moves water from the reservoir behind Stewart Dam to Bear Lake for storage and later release for downstream irrigators.

It became apparent that there needed to be an amendment to the compact, which was allowed every twenty years. Both Utah and Idaho recognized that undeveloped water was flowing into Great Salt Lake, and Idaho was nervous about Utah's plans for development of its senior water rights. If there wasn't some adjustment, there could be a race to develop water, which wasn't in either state's best economic interest.

A third problem raised its head. The compact allowed unrestricted use and development of groundwater. But if wells were sunk above Bear Lake, would that diminish the amount of water available to store in the lake? Idaho thought so and felt its studies showed that groundwater pumping should be subtracted from the allocations to upstream users. But to maintain the senior water rights of users in Box Elder County, the depletion of water due to groundwater pumping would have to be compensated for by UP&L's dwindling share for power generation.

The amendment added a provision that prohibited new storage of water above Bear Lake if it dropped to 5,911 feet. Then a previous reserve of 120,000 acre-feet to the Bear River Migratory Bird Refuge was eliminated with the stipulation that that water would eventually come from Utah's share of the undeveloped water through a letter of understanding with the FWS. That pleased Wyoming and Idaho because they felt that Utah had a more favorable position to apportion water in the lower river. Idaho was given the first right to develop 125,000 acre-feet of water, and Utah was granted second right for 275,000 acre-feet with the last 150,000 acre-feet divided equally between the two states. Utah and Wyoming got the right to store more water above Bear Lake with groundwater depletion accounted for. The compact was amended and signed by President Carter in 1980.

The Bear River Compact essentially sealed the fate of the river, at least in most stretches between Bear Lake and Cutler Reservoir in Bear River Canyon. Not only were the dams riveted to adjacent bedrock, but they were cemented into law. One could point a finger at irrigators or power generators and argue who was most responsible for the dams and the environmental problems they created, but establishing blame wouldn't help the river system.

The dams turned much of the Bear River into a canal. Stretches were channeled to streamline flow, ostensibly to utilize more water for the irrigation/power generation partnership by minimizing seepage. The dams supposedly controlled flooding and made water flow in the river more predictable, increasing efficiency, at least according to human calculations.

But the elimination of flooding had dramatic effects on the geomorphology and ecology of the Bear. Without the dams, flooding would have shown more seasonal variation. By controlling flooding, the dams prevented the river bottom from being scoured in places where silt tended to accumulate; the year-round silt then buried biodiversity. The silt also stayed in the river, rather than being deposited downstream on the historic floodplain or delta. When the river could no longer flood and spread its water and nutrients over the floodplain, it also couldn't create more potholes or refill some oxbows, those biologically rich factories that are wondrously dynamic. When the Bear couldn't swing side to side in its yearly dance, it couldn't replenish the aquifers near it.

By making river flow abnormally high in summer, when it normally might have slowed to a trickle, to move water stored in Bear Lake to irrigators, river management inadvertently kept the fine-grained banks wet, accelerating their sloughing off into the river and adding still more silt. If the banks had been allowed to dry out, willows, bulrushes, and cattails could have grown over the summer and stabilized them for the next spring's floods.

The Bear River Compact came with a price. It enabled the historical stakeholders to provide Intermountain communities with food and electricity more economically by eliminating uncertainty and the expenses of constant legal wrangling. The compact also indelibly inked these shareholders' power into the ledger. But the river and newer stakeholders with fewer legal rights would pay more of the cost.

Jim Boone, editor of Boating the Bear, *does routine maintenance in his boat shop in Lewiston, Utah.*

Boating the Bear

Jim Boone

If there is one name synonymous with boats and the Bear River, it is Jim Boone of Lewiston, Utah. While hundreds of people have floated certain sections of the Bear, Boone is one of the few who has plied his oars on most stretches. For the editor of *Boating the Bear*, the dog-eared bible for canoeists, boating and living on the water have been his passion and center. More than once, they have generated the current that propelled him out of slack-water times in his life.

Boone's boat building is legend in Cache Valley, along with his work with the disabled. A consummate craftsman, he planes his flinty love of boats into every craft he makes. Never having coveted or accumulated much in life, he considers his creations and knowledge of rivers the only gifts he can give to those whose bodies are little more than anchors.

Boone developed a love of water growing up near the west fork of the Whitewater River in Indiana. Being small-town store proprietors, his parents couldn't afford a lot of frills during the Depression, but they were able to come up with five dollars for a week of summer scout camp. Most nights found him on water.

"What was it about rivers that was so emotionally involving for you?" I ask.

"Probably because I did a lot of reading [of James Fenimore Cooper] when I was a kid about sea voyages and river voyages and life in the woods," Boone replies.

"Would you describe boating for you as a sport, or is it something more than that?" I ask.

"Well, it's not a business," Boone answers wryly, although he notes that when he was in the U.S. Navy, it came too close to being a business.

Boone graduated from Annapolis and served during World War II, but he left after his first

Continued on next page

hitch because of the "military nonsense." "We were constantly preparing for something that we never did," he explains. "We were always worried so much about making an impression on the commodore, whoever he might happen to be."

Boone was attracted to the sea but not to the armed forces. Not too long after he left the navy, he found out he "couldn't adapt to civilian life, either," and joined the U.S. Coast Guard as a petty officer. When he realized he wouldn't be promoted, he enlisted again in the navy during the Korean War. Still frustrated, Boone left the military for good when his enlistment ended.

"I didn't want anything more to do with it, ever, ever, ever," Boone says, the repetition an exclamation point from a man of few words. His best times in the navy were when he had a captain who let him maneuver the ship around the piers when they came into a port. "If they had let me do that, I would have stayed in forever," Boone remembers. "I really enjoyed ship handling."

He bounced around, parlaying his Naval Academy education into engineering jobs at Harley-Davidson and technical writing at Thiokol in Brigham City. He also had a stint as a field executive for the Boy Scouts, which introduced him to the satisfactions of volunteering. He tried opening a motorcycle and bicycle business in Arizona and worked with the Phoenix Indian School. When the shop failed to prosper, he went back to school, getting his MA in English from Arizona State University, which stimulated his lifelong love of literature and burgeoning interest in teaching. He took those with him when he fled with his wife Mary to Arkansas in 1972 and joined the back-to-the-land movement.

In Arkansas, the Boones grew enough food to meet their dietary needs and occasionally traded food for something else. Jim and Mary worked on the side to bring in cash, Mary at a business that made oak furniture and a tourist store, and Boone teaching part time in high school and junior high. Later, he thought he had a full-time job in a community/vocational college partnership teaching electromechanical technicians, but it didn't last long. "I didn't get along," Boone says succinctly.

One semester Boone had a chance to teach a survey course in American literature, but it wasn't a good experience. "It was painful," Boone explains. "It was Bible to me, but they [the students] thought it was stupid," especially when he tried to teach them his favorite author, Henry David Thoreau. He eventually left teaching because he couldn't motivate his students and their parents didn't support education.

"The kids were doing their best to forget their heritage and become modernized and consumerized and mechanized," Boone says. They didn't have any basic curiosity in the rich history of their community.

"The older folks there would tell me about how they would hew railroad ties with broadaxes and carry them up to the ridges, where they could sell them for a dollar a piece"—Boone laughs at the insanity of the Herculean task. "The kids and the people wanted to leave all that behind and get the latest machinery and guns and televisions, etc. I didn't want any part of that."

"You must have looked like an oddball to them," I say.

"I was an oddball," Boone replies. "I was trying [to teach] families who had no books in the house."

Frustrated, Boone turned to boats again. This time he began building them with the devotion of a prodigal son, all varieties of boats, each with a different history and special advantage on the Buffalo River in Arkansas. His first was a canvas canoe out of a kit. Next came an eighteen-foot bateau, a flat-bottomed boat narrow at both ends for carrying cargo through rapids. Then came a dory.

Fully absorbed in his reinvigorated passion, he went to a three-week boat school and built

Jean Lown maneuvers a kayak through the fast water in Oneida Narrows.

a twelve-foot skiff. A twenty-foot bateau followed. Then a sixteen-foot skiff, followed by an eight-foot punt, a thirteen-foot bateau, and an eight-foot pram. It almost was becoming an obsession.

In 1986 Boone had had enough of Arkansas. He and Mary effectively decided to retire in Utah after their three kids gravitated west. They rented a place in Cedar City, but after a few weeks, Boone looked at the boats in his backyard and could see grass growing through the seams splitting in the dry climate. It was time to look for a wetter locale. They found Cache Valley, and he discovered the Bear.

Boone accelerated his boat building in Utah. He built a larger punt, an eleven-foot canoe, and a sailing dory. He used them all and sold a few. Boone has floated the three bateaux, both the dories, the sixteen-foot skiff, the canoe, and the punt on the Bear. The only thing that can't be used there is something with too deep a draft.

News of his boat-building prowess began to circulate in northern Utah. That reputation led Allen Stokes from the Bridgerland Audubon Society to contact Boone about writing a book on boating the Bear. Stokes gave Boone incomplete material started by another author. For areas of the Bear that weren't covered by those accounts, Boone filled in the gaps himself by floating and taking notes. He wound up coursing the Bear on various trips from Montpelier to the Bear River Migratory Bird Refuge. These edited notes became *Boating the Bear*.

Continued on next page

Teenagers spend a Friday afternoon rafting on the slow water in Oneida Narrows. "Row like a man," she yells into his ear.

I ask Boone about the changes he has seen on the Bear since he began plying its waters in 1986. "There's been some effort to clean it up," Boone says. "Not nearly enough, I don't think, but there's been some effort."

"So it's better now than it was?" I ask.

"Yeah, not so dirty," Boone replies. "Not so many dead cows on the bank. It used to be horrible."

Boone believes there should be even more cleanup. "They've gotten rid of some of the football field–long banks of manure, but there are still places where cows will be standing out in the water. They just never consider," Boone adds, "that the river is a great recreational access along its whole length. I don't know why that idea is so hard to get across."

Boone points out that, although the banks seem cleaner, there still are old junked cars shoring them up from erosion. They are the remnants of earlier attempts at revetments, a time when farmers used any materials at hand to stabilize the riverbanks, stopgap measures to try to preserve their land before it sloughed off into the river. Boone whimsically calls them "art galleries."

"What about navigation? Are there as many fences across the river as there used to be?" I ask.

Boone says that there aren't too many fences, except a few incomplete ones around Georgetown, Idaho. But that's not the case farther south. There are numerous barbed-wire fences across private property in the stretches downstream from the Forest Service land in the Uintas. Some summer-cabin homeowners are zealous about marking their territory, rights, and privileges. Boaters have been known to carry wire cutters.

Not enough is done to facilitate boating on the Bear, according to Boone. There are marinas at Cutler Reservoir for motorboaters, but self-propelled boaters get short shrift. "I wish that we had more access points," Boone observes. "But nobody seems to care much about it. They only worry about the motorboaters, and the water skiers, and the duck hunters. They build places for them to get on the river."

"All I'd like," Boone continues, "would be just a little access where you could park a car or two or a bike. And maybe modified a bit so that you didn't have to drop down a cliff [to get to the water]."

Boone also would like to see less water-level fluctuation on the Bear and its reservoirs, like Cutler. When boating, "you've got to take into consideration that [PacifiCorp] raises and

A dog serves as figurehead while Catherine Burks and Kate Stephens share a day with Annette Wright. When asked how the river affects her, Wright says, "It's peaceful."

lowers the level of the pond. You don't always know what they are going to do."

"How does that affect you?" I ask.

"Sometimes it's hard to find a channel to get through. Upstream from here, there's a place where, if they really turn it down, you have to know exactly where to go to get through. But I like poking around the river when it's like that," Boone continues. "It's easier to go upstream. It's fun, but if you have a motor, you're really stymied."

Ironically, Boone's dream is to own a motorboat one day. He has thought about the Redwing design, a low-powered, flat-bottomed boat with a small cabin.

When asked if he would automatically launch the motorboat rather than a self-propelled one if he had a choice, Boone considers carefully. "No" is his sly reply, purposely begging another question.

"How come?" I ask, knowing I am becoming his straight man.

"You can travel these waters pretty quickly with a motor, so what would you do the rest of the day?" Boone asks. His years on the river have left him with weathered skin and an equally dry wit.

Common Ground

Common Ground Outdoor Adventures is a nonprofit service agency that sponsors outdoor experiences for persons with disabilities. It gives

Continued on next page

Canoeists enjoy a slower, more-contemplative trip on an early spring day on the Bear River near Trenton, Utah.

them an opportunity to participate in group activities, to observe and learn about the natural environment, and to gain rapport amongst themselves and with able-bodied companions who have knowledge and skills they're eager to share. Jim Boone responded to one of their radio ads soliciting volunteers, forging a connection that would launch a second career as a "boat therapist." When the agency contacted Boone about donating one of his boats for a fund-raising auction, he gave them the sixteen-foot skiff without much hesitation.

"How is the Bear River therapeutic for those clients of Common Ground?" I ask.

"Occasionally I see them trying something they hadn't tried before," Boone replies. "It didn't look very spectacular," he says, referring to one of his charges who tried to row a bit on Oneida Reservoir. "But at least she was trying."

"It gives them confidence just being with people," Boone adds, and it gets them out of the house and away from the television. Boone then describes one of his clients, "Betty," who has limited strength in her hands due to multiple sclerosis.

"If she had to stand up, she'd fall down," Boone says. "But she likes to play with the paddle. She keeps looking back to me and asks, 'Am I doing it right? Am I doing it right?'" She got the chance to be in control from inside a body that doesn't allow a lot of it.

I ask Boone to relate a few other stories about how boating the Bear has been good therapy for clients of Common Ground. Because he is an intensely private man himself, Boone chooses to use fictitious names.

"'Arthur' is a fortyish man with multiple sclerosis," Boone begins. "He believes that without the benefits of a disciplined physical regime, he would now use a wheelchair. However, his regular [boating] activity has given him confidence, and he's largely self-sufficient and able to combat the depression that threatens to become too powerful for one in his condition."

"Charlie" only has partial use of his legs, but with physical determination, he's been able to spend longer periods of time outside his wheelchair. "Much of this improvement is due to exercise on solid ground, but canoeing and swimming are also sources of benefits to his physical health," Boone observes, "and the water-based outings certainly contribute to his making friends." Boone reflects on the pleasure both he and Charlie had one spring when he carried Charlie to an egg-filled goose nest on an island in Cutler Reservoir.

Common Ground has plenty of young, athletic folks on its staff, and Boone now is content to let them do most of that heavy lifting. His youthful comrades revere his skills and experience, and he shares them with them. He still can row, paddle, and set poles with the best of them, but he admits to being slow, usually arriving last at the takeout. Born in 1923, Boone glides to a different pace on the river. His younger companions like to race along with the current. Some days Boone opts to row alone. On other days, he takes on extra passengers. While he sometimes feels that the agency's youthful volunteers, like earlier explorers on the Bear, are too concerned with getting to a place downstream, Boone likes his Common Ground clients to sense the river underneath them. He would be content to let them row in a slow, constant arc to port or starboard—as long as it gives them control.

Irrigation

"Sometimes I feel like the weight of the whole thing is on me," Charles Holmgren says.

As president of the Bear River Canal Company in Box Elder County, Utah, Holmgren is responsible for managing the complex affairs of one of the oldest companies on the river. In good years, there's enough water to go around, and he doesn't have to make Solomon-like decisions. But during times of drought, he needs to insist on water being delivered according to the company's senior rights. That's awkward and an emotional burden because, while the Bear River Canal Company is far downriver from Bear Lake and therefore relatively last in line to divert water from the river, its rights are first in time. So, when Holmgren exercises those rights, it means some upstream user loses irrigation water.

Although Holmgren was born in 1951, his family roots go back to the 1860s, about the time when first-generation western settlers decided that, if they were going to make a success of farming in the desert, they'd have to construct an irrigation system that relied on canals. Bothwell's initial attempts to divert water from Bear River Canyon eventually became the property of U-I, which was a good steward of those rights and faithfully delivered water to the irrigators in Box Elder County. But when the country was racked by inflation during the 1970s, U-I found itself trapped in fixed contracts that required them to deliver water at a financial loss.

"Apparently U-I's attorneys didn't see far enough into the future," Holmgren explains, "and didn't account for inflation. The contracts with Utah-Idaho were for a specified annual assessment. Some people had one-dollar water: a dollar per share per acre."

A drop structure lifts canal water high enough for gravity to allow it to flow through an open headgate.

Typically a farm in Box Elder County uses 4 acre-feet of water per acre during the growing season. This accounts for losses from evaporation.

"When inflation came along and they had these contracts for this one-dollar water," Holmgren continues, "some of it went up to four-dollar water. But they [U-I] were going broke. They had no way of increasing revenue. So one year they decided, 'We can't run the canal system this way. We're going to set an assessment of fifteen dollars an acre.' A lot of the people rebelled. They said, 'I have a contract that says that you are going to deliver me water for one dollar a share.'"

The irrigators filed suit against U-I. The company said that, if it was required to deliver water for one dollar an acre, it would cease to maintain the system. A year could go by before a farmer could get work done on a headgate. The stalemate spurred negotiations by the irrigators to purchase the canal system from U-I.

"They negotiated a purchase price somewhere around thirty dollars per acre," Holmgren says. "A one-time purchase price. It was just a little over a million and a half dollars in purchase price for the whole system. They got nearly 90 percent participation from people in the valley."

A few farmers balked, didn't buy in, and continued to get their water at one or four dollars an acre. The new Bear River Canal Company, formed in 1981, did have the carrot that, if people bought in at the inception, they'd only have to pay thirty dollars per acre. It went to sixty dollars in two years. Now it's ninety dollars an acre.

Headgates on the canal come in a variety of sizes, allowing more or less water to flow through side canals onto adjacent fields or to an entire community like Bear River City to irrigate gardens and lawns.

Then banks started to put inadvertent pressure on the recalcitrant farmers. The holdouts had a contract with a company that technically didn't even exist, so farmers didn't have a negotiable instrument. Without that, banks wouldn't make mortgage loans and allow those old contracts as collateral. Now the canal company has about 99 percent participation of farmers in Bear River Valley.

There are about sixteen hundred shareholders in the canal company, which delivers water to just over sixty-six thousand acres through a system of approximately 130 miles of canals. That water fuels about thirty-seven million dollars in yearly agricultural receipts. The company takes its water out of Cutler Reservoir in Bear River Canyon and then divides it between two canal systems, the East Side and the West Side. The West Side Canal carries approximately 80 percent of the water shares, although the East Side Canal is more expensive to maintain because it lies in the foothills and carries higher liability insurance due to landslides. If farmers want to transfer their water to another shareholder, they have to notify the company.

Because U-I sold its power-generation rights at the Wheelon Dam at Bear River Canyon to UP&L early in the twentieth century in return for guaranteed water, the Bear River Canal Company inherited contractual rights to store water in Bear Lake, and UP&L and its owner, PacifiCorp, are required to manage them. The canal company also has good rights to natural flow in the Bear River that predate almost every other water right on the river. One set of rights dates to 1879. The rights are so important and inviolate that, when Scottish Power purchased PacifiCorp, it had to stipulate, before the Utah Public Service Commission would ratify the deal, that it would honor all rights and that irrigation and flood control were the primary management concerns on the Bear River.

"There's a difference between water rights and contractual rights on storage water," Holmgren explains. "We have some old water rights on the river in the name of Bear River Canal Company. They aren't PacifiCorp's or Utah Power's. They are Bear River Canal Company water rights. We have pretty good water rights up to 333 cubic feet per second."

"So those would be rights where you could put a pipe in the river and pump out water?" I ask.

"Yeah," Holmgren replies. "And every day, as we get into the summer, when we are using Bear Lake water, we still have mixed in with that a third, perhaps, to a half of water will be natural flow. And the Utah state engineer's office has to keep track of what portion is natural flow and what portion is storage water. And that [the latter] goes against us. It's almost like borrowing from your savings account. You have a salary coming in, but you're spending more than your salary, so you're taking it out of savings. You're combining those two sources of money to pay out, and that's kind of how it works with water."

PacifiCorp is obligated to deliver to Bear River Canal Company up to 900 cubic feet per second of water from storage in Bear Lake from May 1 to November 1. Shareholders use that water to grow wheat, the dominant crop, followed by alfalfa, corn for silage and grain, onions, and some barley, oats, sod grass, and mint. But just because the company has rights to that amount of storage water doesn't guarantee that it will receive it. In fact, the canal company has never used more than 124,000 acre-feet of storage water in that six-month period, even though it has rights to more than 327,000 acre-feet. (An acre-foot of water is approximately 326,000 gallons, enough to serve the needs of a family of four for one year.)

The thirstiest crop is alfalfa, even though it has a root about ten to twenty feet long and can draw moisture from the water table. In fact, when the Utah state engineer's office figures out how much water should be apportioned to a farm based on its acreage, it uses alfalfa as the standard because it grows all summer. It requires 4 acre-feet of water per acre. Onions are thirsty, too, because they have shallow roots and spindly leaves.

"Corn has a shallower root. You just about have to irrigate corn in a flood-irrigation situation in this valley at least every two weeks," Holmgren explains. "Wheat can go every three weeks. Alfalfa can go every four or five weeks. Onions need it every week. They are such a larger consumer of water, not that it all goes into the plant but that it goes out across the field and it evaporates. Some days it just about drives me nuts to watch them use as much water as they do for that crop." Holmgren cash-leases some of his land to an onion farmer.

"Generally, if you have a dry spring, you take a lot of water early in May," Holmgren says. "Then, as people get their fields irrigated, we drop back a little bit. Then, in the middle of June to the first part of July, there are crops that haven't needed water, and other crops such as wheat that continue to need water, and our demand really peaks in the middle of June to the first part of July."

"As you get past that early July date, wheat matures, and there's no wheat being irrigated, and flow demand drops off," Holmgren continues. "As you get later into the fall as corn matures, and onions mature, water is cut back until we get to October 1, and then we get into a little bit of a debate with the bird refuge because their water rights in some respects trumps ours. Unless we get a special dispensation from the state engineer's office, we go on nearly exclusively storage water after October 1. Throughout...the fall and winter, we still have what's called a stock water right, where we can run water through the system to water cattle and sheep. On our own farm, we put water out, and it saturates the soil. We have these field drains underneath the ground—six feet under the ground—and that kind of charges the aquifer. That reservoir will carry us through most of the winter to water our cattle."

The Bear River Canal Company delivers the water according to a strict schedule. Every shareholder knows when he or she will receive water. Shareholders have to take it then by opening their headgate to a certain height, in effect, measuring the water by governing the flow. Some small users take a stream of 2.5 cubic feet per second, while Bear River City has a large headgate that gives them a stream of almost 24 cubic feet per second. But if you snooze, you lose.

"It isn't a system like the homes in Bear River City, where you just turn on your tap whenever you want to turn on your tap," Holmgren says. "There's enough capacity in that system that you can do that at any time. In the canal system, we have to balance what's flowing all the time; if you're allowed six hours of water per week, those are the six hours that you have to use. Just the other day, I was in the office, and they said, 'This individual wanted to know if it would be all right. He wanted to leave town for the Memorial Day weekend, and he wanted to know if it was all right to use water on Friday because he wouldn't be here Monday.' We said, 'We just can't accommodate that because you'd be turned on the same time as someone else, and that would drop the bottom of the canal out. It just wouldn't work.'"

It's important to leave just the right amount of water in the canal, too: not too much, and not too little. "As you move water up a canal, away from the lower part of the canal," Holmgren explains,

Charles Holmgren uses an old-fashioned siphon tube to lift water from one of the 130 miles of canals in the Bear River Canal Company system in Box Elder County, Utah, to water his wheat field. The company's water rights date from territorial days.

"you need to leave some water…for a term they call 'carrier water.'" Because even if there's only one individual way down here on the end, there are seepage and evaporation losses along that line. On the other hand, if you are moving water down a canal, you have to be sure the canal has the capacity to handle that move."

Holmgren is a graduate of Utah State University, but the interpersonal skills demanded of the canal company manager and president almost require an advanced degree.

"The worst thing a canal company can do is make the mistake of ever giving somebody a little extra water or missing it in calculation," Holmgren says. "My understanding is that we've been delivering too much water to Corinne City on their particular headgate for the past three years. We discovered the error this year, and now we've put them on their correct stream, and they can't live with it. But they're going to have to," and Holmgren laughs in mock frustration, "because I'm not going to give water to them and take it away from some other shareholder. This is life, and here we are."

The Bear River Compact didn't settle every eventuality, either. In fact, it allows the three states to ignore compact agreements under some conditions and take unilateral action. Holmgren points out one such scenario that occurs when trying to strike a balance between taking natural flow rights versus contract storage ones. Exercising

Migrant workers, overseen by Robert Call, the land lessee, prepare polypipe to draw water from a ditch and irrigate a field of onions. The men will punch holes in the plastic pipe above furrows, creating a sieve for the water that flows through the ditch the next day.

natural-flow rights can force upstream users with junior rights to begin taking their storage from Bear Lake before they want to.

"So they get deducted on their storage water before we get deducted on ours. It gets really pretty complicated. It's so critical to make sure that you're accounting for all this correctly. In fact, we get to the point where we've debated a water-conservation effort to shut the canals off for, say, a two-day period."

"To cut down on evaporation?" I ask.

"No, in order to refill the reservoir at Cutler," Holmgren explains. "We theoretically could shut our canals right off for a two-day period out of every ten, or two out of every seven, or something like that. But if we do that, then we are not taking advantage of our natural-flow water right. So we always have to take a little bit of water, or we are allowing somebody else to take advantage of our natural-flow water right. We have to be very careful that we continue to take water, but not too much, or we'll deplete our allocation at Bear Lake. You have to have all these things going on in your mind to be careful of what you're doing because sometimes you're shooting yourself in the foot, and you don't realize it until a month later, when you say, 'Why did I do that?'"

"One thing that, I guess, it's been impressed upon me through our former manager," Holmgren continues, "is that, from actual experience on the river, we have to be careful to maintain our storage water because, when we run out of storage water at Bear Lake, then the state of Idaho changes the way they operate their accounting model. They've indicated to us that they're going to dry up the

Bear River at the Idaho state line. So, does not only using up Bear Lake storage water affect us from that aspect, but it also hurts our natural-flow water right if Idaho changes from a mutual accounting model to one where they are going to dry up the river at the Idaho state line."

Pressures like this take their toll on Holmgren. "I'm always obsessed with weather forecasts," he says, "but I can't stand to watch them because I'm always disappointed." Even when he gets a rainstorm and would like to stop taking storage water from Bear Lake, he can't.

"The distance from Bear Lake to Cutler is five days from when you do an action at Bear Lake until it gets to Cutler. So you get a rainstorm, and you tell PacifiCorp, 'Hey, you can cut us back,' and they say, 'Well, we can't get this back into Bear Lake. What are we going to do with it?'"

"'Well,' I say, 'turn on your generators and make some more electricity. Otherwise, it's gone.' You wish you could see five days in advance, but you can't."

Managing the stored water at Bear Lake is a touchy situation for Holmgren. "Even though we've had these senior contracts with PacifiCorp," he explains, "we still realize we are at the bottom of the river, speaking of Bear Lake water. We need to work with people on the river to the mutual benefit of everybody. We don't want to hog it all, even though we have this senior contract."

So in 1994 Bear River irrigators formed the Bear River Water Users Association. It is composed of virtually all the water users between Bear Lake and Cutler Reservoir, specifically, the Bear River, West Cache, Cub River, and Last Chance Canal Companies. The association hopes to iron out any differences that may crop up among them and with PacifiCorp, especially during times of drought.

"What is your relationship with PacifiCorp like?" I ask.

"Well, it was better in the past when it was Utah Power and Light because they were in Salt Lake," Holmgren replies. "They've moved so many positions to Portland; from time to time, we get the feeling that 'Hey, they don't know we exist out here.' They're good people, but they are just so far away from us they don't know what's going on. And again, it's a small part of their big pie, the administration of the Bear River. We were really disappointed when they gave Carly Burton an early retirement. We were very happy that they kept him on as a consultant." PacifiCorp hired Connely Baldwin as the company hydrologist in 2004, but it takes years to gain the experience and wisdom to manage the river like Burton.

Relationships between the Bear River Canal Company and homeowner associations at Bear Lake have been strained at times, too, because most irrigation in Bear River Valley in Box Elder County is by flooding. Bear Lake interests feel that, if irrigators switched to sprinklers, there would be fewer problems with lake fluctuations.

"Are your canals concrete lined at this point?" I ask.

"Very few of them are," Holmgren replies. "We're in quite

heavy soil, clay soil." The clay helps prevent seepage, and it's certainly less expensive than constructing concrete canals. Still, it's a pressure point.

"We are chastised, occasionally, by the Bear Lake people for being inefficient," Holmgren says, "but our system versus other systems is really pretty efficient. If we can get across this valley using 4 acre-feet per acre average in a year, we're doing really well. I was looking at a system in the Idaho area that was using 7½ acre-feet per acre. That high number was what was being applied to the crop but also what was leaking through the canal."

"We're constantly encouraged to move over to sprinkler systems," Holmgren continues, "but while we would become a little more efficient capital expense to buy the sprinklers, the pivots, the lines, and things would just be astronomical."

Plus the canal company's electricity costs would dramatically increase because of the need to pump water to pressurize the system. "With a few exceptions, a gravity-flow system is not electrically dependent and not pump dependent," Holmgren explains. "It just works by Mother Nature pretty much. This valley is so flat and has such heavy soil that we don't require a lot of water."

"And then the interesting thing that happens," Holmgren continues, "when you flood irrigate some of that water returns back to the Bear River as return flow. Consequently that's to the benefit of the people below that. It becomes natural flow, even if it had come out of Bear Lake as storage water."

When irrigators upstream in Gem Valley, Idaho, switched to sprinkler irrigation, in part because they lost so much water to the porous, volcanic soil, natural flow in the Bear below that point began to drop off. The downside to return flow is that the water carries pesticides and fertilizers into the river, further lowering its water quality.

Still, farmers are happy to get it, especially during times of drought, like the first five years of the twenty-first century. Less water means lower yield with a concomitant loss in revenue. Farmers try to break the rules if it means they can get more water, even if it's at the expense of another farmer. The Bear River Canal Company ditch riders, who now patrol with cell phones and pickup trucks, rather than horses, have seen just about every kind of subterfuge.

"There are people that will go out and adjust the headgate under the water so that the canal rider can't see it," Holmgren says. "Or put a board in one of these drop structures on one of these dams and raise the water level higher, thereby increasing the pressure on the headgate and pushing the water out of the head gate. These guys are just constantly being surprised with new ideas by 'helpful irrigators.' It's a frustration. We haven't got to the point that we've arrested anybody."

"But you could, I guess?" I ask.

"Yeah, yeah," Holmgren replies. "It hasn't happened here yet, but

Gravity flow, especially for a thirsty crop like onions, causes water loss to evaporation. Yet some of the water drains off the fields and back into the Bear River, where it becomes natural flow for downstream users.

this summer [2003] might be the one that tries people's patience to that point."

Two thousand three was the fifth year of drought in the Bear River Basin. The picture looked dim for all river water users. Holmgren says it looked as if the canal company was going to have to start taking its Bear Lake storage water earlier than it ever had. Its allotment was going to be 15 percent less than 2002's already historically low portion. As it turned out, though, there wasn't too much impact on Bear River Canal shareholders. "We planned well in 2003. We got by. There wasn't much of a change," Holmgren says.

However, 2004 looked like a disaster in the making. In December 2003, Bear Lake was 5,904.1 feet in elevation. If it reached 5,904 feet, the irrigators would be in trouble due to the Bear Lake Settlement Agreement.

"In 1992 the lake was drawn down, and there were sandbars in front of the [Lifton] pumps," Holmgren recalls. "PacifiCorp applied to the Army Corps of Engineers for a permit to dredge in front of the pumps. The residents around Bear Lake—some of them—filed a lawsuit against the Corps of Engineers to prevent them from issuing PacifiCorp a permit. After two years of negotiation with the landowners around Bear Lake, we entered into an agreement called the Bear Lake Settlement Agreement. In exchange for allowing us to dredge, we limited our draw every year—our total draw was unlimited previously; we allowed restrictions to be put upon releases from Bear Lake every year and actually gave up the bottom two feet of water in Bear Lake…for the benefit of the Bear Lake… recovery time, as it was described.

"It is a little bit confusing, that bottom two feet. We can actually use that bottom two feet if the forecast of available storage is above an elevation of 5,904 feet a year. We can pump down to 5,902 feet of elevation. However, if we start out at 5,904 feet or below, then we cannot use any releases from Bear Lake."

Below 5,902 feet, it wouldn't matter because, at that elevation, PacifiCorp's pumps would be high and dry.

Holmgren and the Bear River Canal Company faced a 50 percent reduction in their normal water allotment in 2004. One way the canal shareholders met that forecast was by taking out crop insurance for corn and onions. Of the sixty-five thousand acres in the valley being irrigated, ten thousand went fallow. That amounted to 40 percent of the corn acreage normally farmed. Farmers switched to a larger percentage of cultivated acres for oats and barley. They could get those up with early water and use them for silage or hay for cattle. With those crops, they wouldn't need water in June or July when it likely would be drastically reduced.

The shareholders never had taken out crop insurance before because they had always felt secure with their historical Bear Lake storage rights. They'd never needed it.

"We also defined our water right better with PacifiCorp," Holmgren says. The canal company took its water two weeks earlier, starting around the middle of April instead of May 1. That way they were able to get their canal soaked up. But they never got close to their 900 cubic feet per second of water.

In 2004 60 percent of Utah small irrigators' water was used within the first few weeks. The West Cache Canal Company used up its allotment early in August, and 350 farmers lost their water when their Bear Lake storage ran out. Then the Utah State Division of Water Rights ordered 90 farmers in Cache Valley to stop pumping Bear River water. The state issued the order when the Bear River Canal Company sent a letter requesting regulation of upstream water users.

The Bear River Canal Company had a judgment call to make. Although it had never been done in the history of the Bear River Compact, the company thought about making an interstate call to deal with the water crisis. "That would have been uncharted ground," Holmgren says, and potentially threatening to the compact. Relying on its legal counsel, the canal company decided to make an instate call instead, feeling that would be less portentous. That set in motion the Division of Water Rights' cease and desist order.

Cache Valley farmers directed some hard words at the state and the Bear River Canal Company. The state issued an injunction against one farmer who defied the order, and he sued, although the issue was eventually settled out of court. Still, the loss of water was especially hard for one farmer to digest when his 1880 water right was trumped by the Bear River Canal Company's 1879 right.

Eventually nature helped out in 2004. Farmers got a good rainstorm on Memorial Day that allowed them to shut off release of Bear Lake water for three weeks and rely only on natural flow. They got the wheat crop watered. Then they went on 50 percent water on headgates throughout the system for the rest of the irrigation year, meaning they got half their normal allotment. Holmgren made do with 50 percent less, given his mix of crops. Other shareholders made switches with their neighbors if it was mutually beneficial, based on their crop needs and ease of switching; they took a full ration but only once every two weeks instead of once a week. They also got a good storm around Labor Day that helped carry them through until winter. Even with the help from nature, however, Holmgren personally took a twenty-five-thousand-dollar hit in lost yield, even after a twenty-thousand-dollar insurance payout.

Ironically, when I called Holmgren in May 2005 to see what the prospects were for that water year, he was sitting idly at home, unable to plant some fields or fertilize others because they were too wet from spring rains. Moreover, farmers were having to deal with forced releases of water from Cutler Dam by PacifiCorp, following Federal Energy Regulatory Commission regulations. These abnormal flows in the Bear flooded out some downstream users on the floodplain near Bear River Migratory Bird Refuge. Some of these users asked the canal company to take some of the excess water from Cutler, but Holmgren said they couldn't because landslides from the spring rains had damaged portions of their canals.

In spite of the heavy spring rainfall, the canal company shareholders were contemplating rationing in the 2005 irrigation year, even with a Bear Lake storage runoff forecast 50 percent higher than 2004. It was partly due to a desire to build up the elevation of Bear Lake. It was the result, too, of the maddening vagaries of timing. Even though it had been a wetter winter, in mid-March 2005, the weather got abnormally dry. The U.S. Natural Resources Conservation Service worried about Bear Lake drainage runoff, reporting that it looked as if it was only going to be 60 to 70 percent of normal. So Bear River Valley farmers bought more crop insurance in March. In April and May, the rains came. The farmers were locked in and continued to ration with fallow ground and take the crop-insurance payments.

"I've decided there are only two states of being—mud or dust," Holmgren told me wryly.

Remnant of the concrete "great arch" that once shouldered the flume carrying water from a tunnel, the last engineering problem in building the Last Chance Canal. A utilitarian, but less graceful, structure supports the new flume.

Last Chance Canal

Pioneer Ingenuity and the Need to Claim Water

Earlier attempts to dry farm in northern Gem Valley had failed. The porous, volcanic soil wouldn't hold the sparse moisture. To survive, the settlers needed to irrigate by drawing water from the Bear River. They built a flume by the side of the river, but snow sloughed off the canyon ridges and smashed the wooden carrier to bits. The conclusion was obvious: they needed to build a dam at a higher elevation upstream where they could divert water into a canal and flume that would parallel the ridges.

In 1897 three principals in the dam filed for a water right of 200 cubic feet per second for the Last Chance Irrigation Company. It was an apt name because, if it didn't work, the farmers faced economic ruin. They selected a dam site on the Bear River about two miles below Soda Point. They dislodged boulders from the canyon walls with crowbars and shovels and attempted to build a dam with logs they had cut and floated downstream. But the dam collapsed.

Since they couldn't afford to fail, they tried a second time in 1899 under another name: the Last Chance Canal Company (Ltd.). According to Vivian Simmons and Ruth Varley in *Gems of our Valley,* stock was granted for work in kind or a dollar a share. During the winter, they

Rock cribs that once provided the foundation for a wooden flume carrying irrigation water across the Bear River. The route was abandoned for a later canal built parallel to the rim of Black Canyon.

The Last Chance Canal diversion, approximately two miles below Soda Point. The canal became a monument to pioneers driven by the need to move water to claim a stake on the land.

built cribs on the ice from logs and filled them with lava boulders. In the spring, the cribs sank to the bottom of the river and became the foundation for the dam.

Even with a target of fifty thousand shares, the money ran out. The shareholders had to take out a loan to complete the dam. But to comply with a government ruling, the company's now 440 cubic feet per second of water rights had to be over the "right of way" by February 12, 1902. With eleven hundred feet to go and two weeks to do it, the canal seemed doomed. If they defaulted, the Bothwell Canal Company downstream was ready to file on the water. They would have to build the canal the rest of the way through the snow. They ploughed through the snowbanks, let water trickle down the snow to freeze the sides, and then diverted water from the dam through the ditch, getting the water over the right of way and saving the filing.

It was a stopgap measure. In the spring, they knew they would need to construct a permanent canal, and part of that ditch would have to be built around a rock abutment. But after it was completed, water seeped through the rocks of the ditch wall, and sections of that canal, too, sometimes crumbled into the river and cost money to rebuild; they also lost the irrigation water.

In 1916 the board of directors voted to carve a tunnel out of the abutment to eliminate the need for a canal to encircle it. The board awarded the $12,500 contract to Morrison Knudson. Two brothers began digging through the lava, did much of the drilling by hand, and then hauled out the rock with one horse pulling a small railcar. Starting at opposite ends, they came within one foot of perfect alignment. By digging through that abutment with a 1,476-foot-long tunnel, twelve feet wide and nine feet high, they cut the canal length to about three miles besides making it more efficient.

Next the company constructed a new cross-river flume at the tunnel outlet on top of a concrete structure eventually dubbed "the great arch." In 1947 the board replaced it with a no-frills, rigid steel support, although it kept the concrete arch for sentiment's sake, and perhaps with a nod to the notion that some architecture grows old gracefully.

Cutler Reservoir is a major geographic feature in Cache Valley, Utah. UP&L is beginning to manage the lands adjacent to the reservoir less for farming and more for land conservation and wildlife habitat.

Hydropower Generation

"Hell, normally I would wine and dine you," Kelly Holt, the Idaho hydro manager of PacifiCorp, said when I told him about my Bear River documentary project and interest in using the Grace plant as an exemplar of hydropower generation on the Bear. "But things have changed since September 11."

Any request like mine had to be run through corporate security due to anxieties about terrorist attacks. Power plants are sitting targets. After about five months of negotiations and three different contract rewrites, I reached agreement with corporate security and arranged an interview with Holt. But we could only talk about hydropower operations. Any conversations regarding PacifiCorp's ongoing negotiations for relicensing by the Federal Energy Regulatory Commission for its Idaho operations were off limits. I'd have to speak to still another person in the Portland home office on that subject. It was going to be a tightly controlled interview. No wining and dining.

Cutler Dam and power plant are the last hydropower facilities on the Bear River.

Most of the five different hydropower plants PacifiCorp operates on the Bear River in Idaho are like the one at Soda Point, with a dam, a reservoir behind the dam to build a head, and generators to produce electricity. In the case of Soda Point, the output is 14 megawatts. Oneida works the same way. But the Last Chance/Grace/Cove operation is unique.

Some four miles downstream from Soda Point, where the Last Chance Canal Company diverts its water, the remaining water in the Bear begins to flow through the Grace system.

"What happens," Holt explains, "is during irrigation season, the water comes through Last Chance diversion, comes down the canal to the irrigators' diversion tunnel. During irrigation season, they have rights to the water, to *X* amount of feet. So we generate what's left. The excess overtops their diversion and comes down the river to the Grace forebay. That's where we collect the water and put it into the Grace pipe for the Grace plant generation."

PacifiCorp takes out at that time whatever's not being used for irrigation. "If there is any," Holt adds.

"Are there times when there isn't any extra water coming out?" I ask.

"There are times when there's not enough water to turn even the smallest of our generators," Holt says. There are three generators at Last Chance—250 kilowatt, 500 kilowatt, and 1.2 megawatt—and the bigger the unit, the more water it takes to turn. Sometimes, then, they are not generating electricity at Last Chance. The same thing holds true for the three generators at Grace.

The water collected at the Grace Dam is funneled into a five-mile-long wood-and-steel pipe with a diameter of eleven feet, six inches. The pipe is always full unless maintenance work requires dewatering. If there is any extra water, it flows over the Grace Dam

Water fluctuation levels below Soda Point Dam are dramatic.

The Bear River below Cutler Dam and Bear River Canyon shrinks in size since the Bear River Canal Company diverts much of the water for its East Side and West Side Canals.

and down through Black Canyon to the Grace power plant. But springs provide the principal, constant flow in Black Canyon.

There's less than a 50-foot drop from the Last Chance diversion to the Grace Dam. There's an 80– to 100-foot drop from the Grace Dam to the surge tower. There's a big 450-foot drop from the surge tower through the penstocks to the generators at the Grace power plant. The Grace plant, in turn, can generate 33 megawatts, 11 megawatts from each generator.

Once the water leaves the Grace plant generators, it goes to a holding pond, the Cove Reservoir, and at the bottom of the reservoir, it enters a wooden flume for the last mile to the Cove plant and through its metal penstock. Cove can generate 7 megawatts with its single generator. At Cove, the water goes back into the main stem of the Bear River. From the Grace plant to Cove, the only water in the river comes from flows that exceed the Cove plant's generating capability, seepage, and springs.

"How do you coordinate with Soda and Oneida?" I ask. "Is each one an independent operation?"

"Depending upon how much water is coming down the river, I and my foremen and operators at each plant will discuss how much water we have coming. If it's during irrigation, we'll discuss what the irrigators need—like at the Last Chance Canal: how much will they be taking, and what's left to get through Grace to get down-river to the irrigators below here? So there's a lot of phone conversation between the plants."

When asked if the water PacifiCorp stores in Bear Lake is water that Holt moves down the river to generate power, he responds quickly. "No. The only time Utah Power and Light/PacifiCorp uses the water out of Bear Lake is for two reasons. One is for flood control. If the lake is too high in the fall to accept enough spring runoff, then we'll generate through the winter with that water to draw the lake down in order to catch the runoff so we don't have flood conditions down below Bear Lake. The other time we would use the water [in Bear Lake] is during irrigation when it's coming past us. But," and here Holt adds some emphasis to his otherwise matter-of-fact delivery, "we do not use any Bear Lake water for power generation solely anymore."

"Is it a relief at times to have Soda Point since it seems to be a standard operation?" I ask.

"The only thing that's constant is inconsistency," Holt replies, "because depending upon what's coming out of Bear Lake and down through the river to Soda dictates what they put out. But then you have to temper that with what the irrigators want during irrigation season. So you may be holding a little bit in Soda until the irrigators need it, to try to keep a full pond to meet their needs. So there's nothing consistent about the Bear River."

"Is it safe to say, then, that the Bear River is much more of a sophisticated and complex operation than most other rivers in the West that have hydro because of irrigation?" I ask.

There are three generators at UP&L's Last Chance power plant, but when the Last Chance Canal Company takes out its allotment upstream during the height of the irrigation season, there isn't enough water left to turn the turbines. The only flow in the Bear below Last Chance comes from seepage and springs.

"I would say that," Holt replies. "I don't know if anybody would dispute it, but I don't think you can get any more complex than we are."

Having to manage hydropower operations around irrigation needs is unique and a challenge for Holt. It also forces him to wade into long-standing water-rights and the historical backwaters of water wars.

"It's amazing," Holt observes. "When I first came into this country and into this job, I knew that water was important. But over the past four or five years, I see why they had range wars over water, and people got killed for water. It's almost that important today." He emphasizes the last word, reflecting the contemporary situation. "And to think about what it was like back then. It's amazing. Yes, the uniqueness on the Bear is irrigation."

"Do even small changes require a lot of negotiation?" I ask.

"Oh, absolutely," Holt replies, "because the irrigators own the rights, or PacifiCorp owns the rights with contractual obligations to irrigators, depending upon which irrigation company it is. Very complex."

"And depending upon which irrigation company predates PacifiCorp's rights versus PacifiCorp's rights predating other irrigation rights, yes?" I add.

Hydropower generation on the Bear once accounted for 75 percent of UP&L's system capacity. Now that the company is a part of PacifiCorp, Bear hydropower plays a smaller role, but electricity demands in the West continue to accelerate with increasing population and development.

"Yeah, again, we may have rights, but we have contracts subjugating those or that water right at certain times of the year," Hold responds.

It is getting late in the afternoon, close to quitting time, and Holt and his foreman, Claude Sant, are anxious to get home. Sant has a small spread in Gem Valley, where he raises some hay and cattle that need to be tended before nightfall. Holt, the former mayor of Grace, usually puts in a few more hours on community and county issues before dinner. They are fully integrated into the agrarian community.

It has become obvious, too, during the interview that water has such a tortured history in the West that people in the water business shield themselves with a "let sleeping dogs lie" stance. Most of the water wars have been settled and have locked-in rights—and privileges—stamped with the seal of the history of the area. Families and water intermix and become consecrated. That can produce a powerful alliance of silence. Even though water law has made those agreements iron clad and wrapped them in a mass of Byzantine knots, the emotions triggered by water still rest on the surface. It's as if the wrong word will rip through the thin veneer that covers the emotional scars and let the pain leak through again.

Ashley Wiser from Jack Greene's environmental-science class collects macroinvertebrates in the Bear River Bottoms.

Field Laboratory on the Bear

The River Has a Lesson to Tell

"I think it's critical to get students reconnected to this very essential part of what they are—their local and regional landscape," Jack Greene says. "How incredible it is. Not just from a survival point of knowing how it works, and what we're doing to it and how to restore it, but also just the pure enjoyment. Some of my fondest moments are with my students in the field and these wonderful instances when they say, 'Ah! That's incredible!' They'll never forget them, and neither will I. They're life changers."

Greene, an environmental science and biology teacher at Logan High School, is passionate about the environment and uses the Bear River Basin as a field laboratory. He lets the river teach his students, cannily knowing that its lessons will be indelible.

Greene always tries to link students with working professionals in the field so that they can get a feel of what their life's work may be. Teaming up with instructors in Utah State University's extension division, his students have done restoration work on particularly damaged stretches of the Bear, eradicating nonnative tamarisk and salt cedar, an invasive species that's replacing a lot of willow and cottonwood. They've monitored streams and wetlands. They've pulled noxious weeds.

Continued on next page

Jack Greene and his Logan High School students, Jenna Forsyth and Erika Neely, count birds as part of their field project monitoring the health of the Bear River.

They've planted native trees and shrubs on riverbanks that have been denuded by grazing. His students have also erected some fencing to keep cows off sensitive areas and done GPS mapping to locate problem areas.

"Could you characterize what your students' feelings are about the Bear River before they get introduced to it? Is it something that's just there, that they haven't thought much about?" I ask.

"Oh, absolutely," Greene says, reflecting on what it is to be a teenager. "You are mainly interested in your friends and, of course, having a good time. Going to the mall. Your computer and video games, etc. A lot of my colleagues are saying this: that students are getting further and further away from the landscape that they are so dependent upon but don't realize it. Plus they come out of high school knowing very little of what's out there. They don't know one tree from another, one bird from another."

Greene is concerned that high school biology departments tend to reflect the attitudes toward natural science that exist in university departments because that's where today's teachers come from. Ecology and natural history have taken a backseat to genetics.

"[High school students] may know a little bit of molecular biology," Greene observes. "We're pretty strong there. Genetics, etc. But as far as how things work in the field in an ecosystem, the ecology, the interrelationships, very little of that, and especially in their own area. They don't know their own bioregion, their own home place. So I really focus on that."

Greene has a complex, personal history with the outdoors. Born in 1942, he grew up on a dairy farm in rural Wisconsin and characterizes his childhood as "roughing it." He was a consumptive wildlife user as a child, loving to hunt with his father, uncles, friends—anyone who would go with him. But as he grew older, he became more a student of the environment. "I became a naturalist, an ecologist. Then, eventually, educator, although that was not in the original plan at all," Greene summarizes.

His bachelor's degree is in chemistry and biology. When it came time to get a master's degree at Central Michigan University, Greene says, "I was actually hoping to head into the national parks as a teacher/naturalist. So I took a program that was a dual major: outdoor education and field biology."

Greene was working at the Ogden Nature Center at the time and took the position as its first director/naturalist when he graduated. Since then, he's taught at ten different institutions, both at the college and high school levels. In between, he also worked at the Utah State Industrial School, where he organized an Explorer post.

"I had these hardened criminal kids being Boy Scouts," Greene remembers with a smile. "The reason I did it was so I could take them off campus and do field experiences. The administration forewarned me. They said, 'Hey, you take these guys off campus, and you'll never see half of them again.' But in all the outings we had, I only had one student run on me. The others—they were cool."

Greene feels fortunate that he has always had administrators who have given him the latitude to take his students into the field. That's been important because he uses environmental activism as a pedagogical tool.

"My purpose as a educator," Greene explains, "is, one, to empower my students; two, to make them all naturalists; and three to make them all activists because our democracy, quite frankly, is falling flat on its ass because very few people are active anymore because they don't have time or they are disillusioned, or [they ask] 'What difference can I make?' Well, I want to convince them that they can make a hell of a difference. And we have. I can cite a number of instances where we've made a real difference. The power of one, so to speak, and especially the power of the small, dedicated, informed citizen."

Greene's environmental-science class is more than just field biology. A wide-ranging thinker and well read, Greene looks at field ecology from a macro point of view. During the course, he weaves in topics from economics, sociology, philosophy, and ethics. In short, his classes are an integrated curriculum, largely because he maintains that human beings are part of the ecological systems and that that needs to be appreciated. He's a passionate believer in sustainable development and challenges his students to explore that problem.

"That's the big question I pose to them," Green explains. "How can we have a healthy economy in perpetuity as well as a healthy environment? We really start rolling up our sleeves and working on that. We start looking at things like natural capitalism, that wonderful piece of work that was written and published a few years ago by Paul Hawkin and Amory Lovins. [There is] a lot of very sound, well-researched data that shows that we can have both. And it's been going on in a number of countries outside our own."

When asked if our own country is making headway, Greene is unequivocal. "Our own? Oh no, it's way off base!" he says sharply. "But there's a number of European countries—New Zealand, Australia to a degree, Germany, definitely, and so on. They have actually taken on a lot of these new systems of economies and political systems that have allowed their people to be very well served educationally and have all the goodies but do it in a far less damaging manner by basically not having a linear system where you use it and throw it away. They have closed loop systems where there are no more landfills; there's no more waste. There's very few toxics. You can design around these things. In fact, the nice bottom line is that those economies are actually providing more jobs per dollar of input than is ours. So their economies are doing very well by making these transitions."

Aware of the oft-heard charge that environmentalists are elitists, Greene teaches his students to avoid sounding like a plaintive cry from the wilderness. Instead, he instructs them on ways to be proactive. He admires

Continued on next page

organizations like the Nature Conservancy because they look at the larger picture to get things done on the ground. He highlights the Conservancy's Farmland Trust, in particular, because it is preserving open space and sustainable agriculture at the same time as it argues for moving away from farming monocultures and toward more biodiversity.

"We have to work with the whole system here," Greene explains. "We really look at sustainable economics, sustainable businesses, greening businesses, earth-friendly businesses. We look at how we can change our campaign systems so that we can actually have individuals involved in our political and policy-making institutions that don't have these tremendous biases. Eighty-three percent of all our campaign financing on the federal level comes from one-tenth of 1 percent of our citizens. That is not a representative government. And I think you can find pretty much the same quotes on the state level. So getting involved in these things and starting to make these changes is just critical for us, our landscapes, our environment, our future."

Greene also is aware of other political challenges of environmentalism. In spite of the broad agreement for environmental protection in the electorate of both red and blue states, it's as if the lessons of the 1970s have to be taught again to another generation of Americans—and especially to their political leaders.

"I'm really after the big picture here," Greene says. "To know this stuff and appreciate it is great, but you need to take it to the next step. You have to make systemic changes in our economic systems, in our political systems, in our culture. You have to make these changes, or we're just headed to a very difficult, difficult time in the not-too-distant future. I think I can speak with authority because I've been at this game for about thirty years now. I'm really closely linked to those who really are in the know far more than myself, who spend all their time studying birds in tropical rain forests or economic systems as they enhance our environment."

"There's a lack of political will," Greene continues. "Lack of political leadership. There's very little of that. Unfortunately, the good people who try it don't last long. They don't survive. Because you have to be a team player in Washington, or you're not successful. Well, that's what most of them think, and maybe that's true to an extent, but there have been those who have more or less stood their ground and had some success in moving in a different direction that's not self-destructive. Yes, there's definitely hope out there."

The Bear River is that object of hope for some of Greene's students. It has become the grail for their environmental crusades. One political issue that seized their passion not long ago was the proposal for another dam in Oneida Narrows, one that would drown another four miles of prime riparian and wildlife habitat, as well as a potential blue-ribbon trout stream.

"I let my students know that this was an issue and if they were interested, they had an opportunity to speak their minds at a public hearing in Preston, Idaho. So they made up a bunch of signs: 'Save the Bear River!' 'Damn the Dams!' The Federal Energy Regulatory Commission, who was making the decision on this, took their posters as public comment and took them back to Washington. They thought that was really cool. We actually made it into the book."

Many of the student activists sign up for Greene's LEAF program, an extracurricular activity. LEAF, the acronym for Logan Environmental Action Force, allows students to get involved in a variety of different school and community service projects that tie in with a quality environment. But Greene has to be careful how he approaches environmental activism in politically conservative Utah. He can inform, but he can't proselytize.

David Leucke and Ashley Wiser collaborate with Brian Nicholson from the Utah Division of Wildlife Resources. Greene makes sure his students work with professionals in the field so they can "try on" a possible vocation.

"We have to be careful about how we do this," Greene begins slowly. "I'm not in there to indoctrinate. I'm in there to base decisions on good, sound science because that's my perspective. I might tie in some ethics with that science," Greene chuckles slightly. "I'm very much into ethics, although I'm not an organized-religion person. I'm very much a believer in the sacredness of life. I think it's miraculous, okay, from the smallest organism to the most complex. So what I do is, I offer them opportunities to get involved in issues. I let them know right up front that my bias is that I'm an activist. They know that. They read my articles."

A reader for advanced placement tests in environmental science, Greene has also become active in a national student program called the Envirothon, where teams of high school students come together and compete on natural-resources management issues, like forestry, wildlife, soils, water, and aquatics.

"I started pushing for something like that in Utah because we didn't have it," Greene says. "I started talking around and got a fellow teacher and colleague in another school who has similar interests to my own, and we pushed, and we got the whole thing going as far as state competition. We just finished our sixth year, and I have a number of teams. Logan High teams have won four years in a row—the state championship. Which means they compete in the international competition. I can say that every one of those students who's competed and won at the state level and gone to nationals has pursued a career related to natural-resources management, even though when they started doing this, they didn't know where they were headed."

"You were saying that you are rather optimistic about the Bear River. That you feel like it's on the upswing rather than on the downswing, from what you've seen over the past fifteen, twenty years?" I ask.

"Yes, it is." Greene responds. "And I say this because I, again, have a lot of professional colleagues in the NRCS, the Natural Resource Conservation Service, with USU extension, with the UACD, the Utah Association of Conservation Districts, who are doing very

Continued on next page

good things with the agricultural interests. Over ninety percent of the damage to the Bear River has been from agriculture, and that's true with most streams and rivers wherever you go. It's agriculture. They dewater it. They pollute it. They divert it. They channelize it. They drain wetlands and such.

"But all of these interests I mentioned are really working closely with the agricultural interests, and in most cases, have been very successful. There's money available. There's either grants or long-term, low-interest loans, matching grants, and so on, as well as other subsidies that will help these folks come up with good plans and get funding to bring about these sustainable practices. Of course, a lot of these tie in directly with the health of the Bear River."

"The Bear River," Greene continues, "is like the canary in the mine. If the Bear River is healthy, so is the land around it. That means we are doing some good things. And it is turning in that direction." Greene's voice then becomes more excited. "And, also, I should mention PacifiCorp's Eve Davies, the biologist, has worked extensively with the agricultural interests."

"Do you think Cache Valley people are starting to become more aware of the Bear and how it's threatened, in a sense that they see it as a threat to them and their overall well-being, even if they are not attached to the river?" I ask.

"Very gradually. Of course, I do it in my classes, but my students represent just one-tenth of 1 percent of the population of Cache Valley," Greene responds. "And also, we've started this wonderful thing called the Bear River Watershed Celebration. We have a big to-do out at Willow Park in Logan, where we have all kinds of agency and organization people who come in and provide a bunch of information and fun activities for families to get them to go out and get acquainted with this marvelous resource: the Bear River watershed. Certainly, it's improving—the awareness—but we have a long way to go."

For Greene, it always comes back to science: basing environmental policy on the best science, and not placing politics above that science; and not using the margin of error that is inherent in all investigations as a political rationale for maintaining the status quo. Environmental failure is too great a risk.

"So watch what the hell you are doing," Greene says slyly. "Know what you are doing, man. Know the science. And there is a lot of unknowns, but there is enough known that it's a serious concern."

Greene doesn't believe enough work at the appropriate time is being done in environmental science. The first impulse of society is to build a dam and do the science afterward. "Very short-term measures with long-term problems of misuse," Greene observes. "And 90 percent of the time, political decisions as opposed to sound science decisions."

"That's the way it goes," I respond.

"That's who we are," Greene concludes.

Epilogue

Jack Greene retired from teaching at Logan High School in 2005. He's still actively involved with students and Bear River field projects through the LEAF program and by advising participants in the Envirothon competition. He has also forged a partnership with the Department of Environment and Society at Utah State University to become a satellite site for Bioneers, an organization committed to restoring ecosystems to health using the best environmental science. College and high school students in Bioneers team up with Greene for environmental projects on the Bear and elsewhere in the watershed.

The Evanston, Wyoming, Lions Club fishing derby on Sulphur Creek Reservoir. Typically fourteen hundred fishers sign up for the fund-raising event.

Winter Sports in the Bear River Basin

Winter in the Bear River Basin can be hard and long, especially in those areas on the east slopes of mountain ranges where temperatures plummet. To live in winter in spite of it, or to accept the season's dare, hardy residents can bundle up and cart their snowmobiles into the Uinta Mountains or take to the ice.

In January, the Lions Club of Evanston, Wyoming, snubs the biting winds and holds its annual fund-raising fishing derby at Sulphur Creek Reservoir on a tributary of the Bear. The derby began in 1994, largely due to the efforts of Bill Gray, who was looking for a fund-raising idea for the Lions Club. He envisioned an ice-fishing derby that would counteract some of the general complaints about these events. He felt a derby could be successful if it was *controlled*. People could come on the ice with their snowmobiles, other vehicles, or sleds the day before the

Continued on next page

Making the most of winter on the ice ponds in Evanston as enthusiasm for the 2002 Winter Olympics spreads throughout the Bear River Basin.

derby and set up their operations (ice-fishing sheds, tables, heaters). But by controlling access to the lake and posting guards at night, event planners could ensure participants that their gear wouldn't be stolen. Control also would mean constant monitoring by volunteers from the Evanston/Uinta County Search and Rescue units. Gray said they hadn't had any thefts in eight years.

At 8:00 a.m. on January 19, the opening day of the 2002 derby, the temperature was seventeen degrees with a light breeze. Eleven hundred people had registered and paid the thirty-five-dollar registration fee. Gray felt sure they would reach fourteen hundred registrants, the so-called safe load limit decreed by the Wyoming Game and Fish Commission, because of past success. The first year of the derby, Gray and others had warned the area motels and restaurants to expect a lot of people. They knew there were folks flying in from as far away as Alaska. But business owners didn't believe the estimated numbers, and the restaurants ran out of food. The derby grossed more than $700,000. In a typical year, the Lions Club nets $40,000 for charity.

To appease the locals, who initially feared the derby frenzy would deplete the fish in Sulphur Creek Reservoir, catch-and-release fishing is encouraged but not required. A six-fish limit is in effect during the competition. When a person catches a fish, he or she waves a blaze orange baseball cap, which comes as part of the entry fee. A spotter races to the site on a snowmobile, measures the fish, and encourages its release. Gray says they have an 87 percent return rate, although crimping barbs isn't required, and most fishers use bait or artificial lures with jigs.

Fishing was slow that morning, with Gray theorizing that it had to do with the overcast skies. Fish are sensitive to light levels, he said, even through the snow and ice, and during one brief moment when the sun came out for about ten minutes, there was a flurry of people catching fish. The winning fish typically ran from seventeen to eighteen inches long, and the longest fish in all three hourly segments from 8:00 to 11:00 a.m. was eighteen inches.

In another gesture of self-control, the Lions Club takes out a one-million-dollar liability insurance policy, which means it can't serve alcohol. However, people can bring their own, and on that gray winter day in 2002, there were several drunks by 9:00 a.m.

Derby prizes were generous. Seven fish had been tagged and released into the reservoir: five fish with $2,000 tags, one with $10,000, and one with $25,000. In addition, there were hourly prizes for the longest fish, no matter the species. (The reservoir holds rainbows, cutthroats, brook trout, and a few German browns that have worked their way in.) The prizes were one hundred dollars, fifty dollars, and twenty-five dollars for the three longest fish caught each hour in three categories: regular fishermen, women, and youth. There were also various "shore prizes" for best outfit, best creel—almost anything generated a prize. The reason wasn't important. The idea was to keep interest high, even as numbness set in.

Honeyville in Box Elder County is on the short list of preferred dam sites. A reservoir would inundate valuable farmland and Hansen Park during late spring but would turn to cocklebur-infested mudflats after irrigators drew down the water in late summer.

Old fights New
Borders shift
Numbers grow
The slope steepening
By the power of many.

You crowd my banks
Draining your refuse,

Cloud my currents
Proclaiming they cleanse themselves.

Unused water,
Wasted
You say
We need another dam.

Drought looms
For my nest
And your soul.

(left) Agricultural enterprises are the primary cause of pollution of the Bear River system. Private cattle feedlots built in the 1930s still cluster next to the river, many out of sight of the general public. Manure invariably washes into the river.

A dam and reservoir at Amalga/Barrens would cover Clay Slough, an arm of Cutler Reservoir, as well as fourteen hundred acres of valuable wetlands and mudflats that would be difficult to replace.

Chapter 6

Damn the Dams: Conflicts Roil the Bear River

The air was oddly electric for a Jordan Valley Water Conservancy District (JVWCD) meeting in West Jordan, Utah. The board of trustees of the largest water wholesaler in booming southern Salt Lake County was convening in June 2000 to approve the budget—a prosaic subject that normally would attract only a cub reporter for a weekly newspaper—but everyone in the room knew that damming the Bear River was going to be the real topic of conversation. The Utah Rivers Council (URC) was bringing a busload of river activists to protest the dams, and northern Utah ranchers had arrived early to protect their land and livelihoods.

Earlier that spring, to consolidate the opposition, the URC had held public-information meetings in Hansen Park in Elwood, Utah, a lovely green spot that would be inundated if the Honeyville Dam was built. One by one, a string of speakers representing various northern Utah interests told people gathered in the park why damming the Bear was a bad idea.

Al Trout, manager of the Bear River Migratory Bird Refuge, said it would deprive the world-renowned marsh of 20 percent of its water. Greg Iverson, the mayor of Elwood, said the reservoir would back up into town basements via groundwater, and probably into septic systems. A 110-year-old irrigation and drainage system put in place by Box Elder County pioneers to make the land cultivable, and which had received an award from the 1893 Columbian Exposition, would be infiltrated.

The mayor of Deweyville, a town just upstream from Elwood that also would lose farmland if Honeyville was built, said, "We want to maintain our rural lifestyle. We want to keep things the way they are."

Patty Timbimboo-Madsen, cultural resources manager for the Northwestern Band of the Shoshone Nation, talked about how the reservoir would drown the remains of her descendants buried along the river's banks. In a larger sense, her theme was really the ongoing displacement of her tribe. She pointed out that only 470 Northwestern Shoshones were left. The tribe had lost almost that many in the Bear River Massacre, and she worried about how much longer it could maintain its bloodline. The Northwestern

Shoshones can't marry much more within the band due to close family relationships. So they have to marry outside the tribe.

"Now we have Shoshones with blond hair and blue eyes," she said, and they are down to an eighth Native American as a definition for tribal membership.

So tensions were understandably high at the water-development meeting, and Paul Henderson, JVWCD board chairman, sensed it as he nervously tested the timer that would hold speakers to two minutes each. It was standing room only in the small auditorium, and people overflowed into the foyer. Residents from Box Elder and Cache Counties had brought a fifty-foot-long laminated banner that stitched together heartfelt messages from folks trying to get the board to understand how a dam and reservoir would drown their family histories and dreams.

Before Henderson began the meeting, I overhead two water developers assessing the issue. Referring to the Bear River and the Bear River Migratory Bird Refuge, the representative of the Hunter-Granger Improvement District in Salt Lake County said, "There's a lot of water there." The spokesperson for the Utah Water Users Association replied, "Yeah. Over 1.2 million [acre] feet go into the refuge every year that isn't being used." That position rippled through the seats occupied by the water developers, and the phrase, "water going into the Great Salt Lake is wasted," was oft repeated.

The meeting began with a staffer offering a PowerPoint presentation on the next year's budget. No one paid much attention, even though it included a figure for purchasing rights-of-way to bring Bear River water south to the Wasatch Front metropolis. Immediately after the slide presentation and before the public-comment period, a trustee asked the digital presenter if there was a Bear River dam in the plan to divert 100,000 acre-feet of water to Davis and Salt Lake Counties. The presenter said, "No." That trustee then asked, as a means of making a point, if a representative of the Utah State Division of Water Resources had indeed said at the previous year's budget hearing that there was no need to build a dam at Honeyville or in the Amalga/Barrens area to divert the water; that the state would build a dam only if the farmers and ranchers in Cache and Box Elder Counties wanted it to secure their untapped water rights in the Bear River. The presenter said that was correct.

Then the questioning trustee rhetorically spoke to the chair, and the audience, about the upcoming public hearing and what he felt was an unnecessary discussion. He reiterated that the JVWCD was not going to build any dams, and furthermore, the state had said that it would not do so, either. He said it wouldn't happen and that he didn't want to see a dam built, either. Paul Henderson then apologized to all the people who had driven so far from northern Utah to attend the meeting, feeling they had to protest a dam that would inundate them. Other trustees said that even if a dam was going to be built, it wouldn't be the responsibility of the JVWCD, even though the agency knew it would be one of the recipients of

At the June 2000 budget meeting of the Jordan Valley Water Conservancy District, a coalition of northern Utah ranchers and farmers and urban river activists unroll a fifty-foot-long banner of appeals from residents, including families who would lose homes and land they worked for generations if Utah decided to build a dam on the Bear near Honeyville in Box Elder County.

the water. Rather, it would be the task of the state of Utah. It was if the trustees were trying to wash their hands of the matter, not acknowledging that the water wholesaler's thirsty customers were driving the need to develop Bear River water.

That bit of gamesmanship confused the issue. Some of the first speakers, ranchers and farmers, were somewhat reticent when they came to the microphone, partly because they weren't protesters at heart, and partly because they had come thinking dams were going to be built. They apologized to the board of trustees for their confusion but reiterated that they would not like to see dams constructed in the future. Arthur Douglas, president of the Utah Farmers Union, said his organization didn't want to see a dam and land inundated but wouldn't mind the water being diverted if Salt Lake Valley needed to exercise its rights. Several of the ranchers took the same tack, some wryly acknowledging that Salt Lake Valley must need every bit of water it could find if the "mind-boggling" construction they had witnessed driving down was any indication.

But younger activists didn't feel any diminished passion when they took their turns at the mike. They argued that Zach Frankel, then executive director of the URC, should have more than two minutes to talk since he represented so many concerned citizens packing the foyer. When Henderson said Frankel could only have his two minutes, the activist speakers extemporized and moved away from protesting dams to other charged issues. They keyed on elements that had been stated or implied in the budget presentation: purchasing easements for the pipeline; the threat to the Bear River Migratory Bird Refuge and the possibility of dewatering a thousand of its critical acres; and the disgraceful lack of real conservation measures practiced by residents of Salt Lake Valley or adequately promoted by the JVWCD.

Among the speakers, the most effective were the ranchers and older activists from Cache and Box Elder Counties. Often they were one and the same. They approached the board with a degree of respect, not wanting to appear like rabble-rousers, just concerned local folks, and the board members nodded and seemed to listen more intently. The older speakers, through experience no doubt, knew that you capture more flies with honey than vinegar.

The younger speakers lacked that kind of public-speaking finesse—or life experience. They tended to be strident and accusatory in their remarks, suggesting that the board of trustees was the problem, when in fact the board was just reflecting the consumptive self-interests of many people in the West, some who try to avoid the reality that water sources are finite. The trustees tended to tune out the youthful activists, not asking them any questions. But the young people had their eye on a distant future in their lifetimes, and they would not be denied.

Bear River Development Act

Tensions over the Bear River erupted in the first years of the twenty-first century for two reasons: drought and an action taken by the Utah State Legislature in the last decade of the twentieth century. In 1991 Utah passed the Bear River Development Act. It directs the state's Division of Water Resources to "plan, construct, own and operate reservoirs and facilities on the river and to market the developed water."

Ever since the passage of the Bear River Compact and its 1980 amendments, Utah had been looking at ways to use its allotted water. The amended compact gave Idaho the first right to allocate 125,000 acre-feet of water. (An acre-foot is approximately 326,000 gallons, historically enough to supply a family of four for a year.) Utah was allocated a second right for 275,000, and each state received a third right of 75,000 acre-feet apiece. Since Salt Lake County's water demand was forecasted at that time to exceed the supply by 2015, Utah was anxious to develop its share.

The water would be subtracted from the average 1.2 million acre-feet inflow of the Bear into Great Salt Lake, although that inflow has varied widely, from 375,000 acre-feet in 1934 to more than 3.6 million acre-feet in 1984. The amended Bear River Compact assumed there would be 275,000 acre-feet of developable water at Corinne after Idaho and Wyoming took their shares, so the Development Act charged the Division of Water Resources to develop 220,000 acre-feet of undeveloped water-right applications

The Barrens area near Amalga in Cache Valley got its name because of its supposedly barren land. An off-stream reservoir there would require pumping water into the depression.

held by the Board of Water Resources. No agricultural water rights needed to be purchased. The act also stipulated that the division could not begin construction of any project until it had sold or leased at least 70 percent of the developed water.

The Bureau of Reclamation had looked at potential dam sites on the Bear River since the 1950s when the compact was being negotiated. Those studies were updated in the 1970s. During the high precipitation years of the early 1980s, when a rising Great Salt Lake threatened millions of dollars of property damage to railroads and mineral facilities, Utah looked at dams on the Bear as a way to control the level of the lake. When the state experienced low rainfall in the 1990s, momentum again grew to dam the Bear, not for flood control but for reserve. A kind of perfect storm in water development seemed to be brewing.

The Development Act divided the water among four entities. The Jordan Valley and Weber Basin Water Conservancy Districts would receive 50,000 acre-feet each. The Bear River Water Conservancy District would have 60,000 acre-feet for Box Elder consumption, and Cache County would be allocated the remaining 60,000 acre-feet. Moreover, that Bear River water would be used primarily for domestic purposes, mostly along the Wasatch Front. Given the historical percentages of water use, that effectively meant that the diverted water would be sprinkled on Kentucky bluegrass. The

Raising the dam fifty feet at Hyrum Reservoir would enable the Bear River Migratory Bird Refuge to convert a water-flow right to a storage one so that the refuge could receive water when it desperately needs it at the end of the summer. A taller dam and enlarged reservoir would require moving the campground at Hyrum Reservoir State Park to a higher point on the other side of the reservoir.

1992 Utah State Water Plan for the Bear River Basin estimated that the costs for building dams and reservoirs and wildlife and wetlands mitigation was $270 million, an arguably low figure. Other costs, like aqueducts, rights-of-way, and other infrastructure, would be borne by the water purveyors, who also had the ability to levy property taxes. The state also figured that full implementation of the plan would take thirty to thirty-five years. Construction costs would be repaid at 6 percent interest in no more than fifty years.

A loaded question remained. Where would the dams be built? What would be their environmental, social, and economic costs?

Initially the Division of Water Resources located and evaluated almost forty sites, using those previous surveys. Some were more feasible than others after considering variables such as reservoir capacity, stream hydrology, water quality, and environmental impact. Eventually the division narrowed the list to six sites: Honeyville in Box Elder County; Avon, Mill Creek, and Amalga/Barrens in Cache County; another dam and enlarged reservoir in Oneida Narrows; and an enlargement of Hyrum Reservoir by raising the dam fifty feet. The last site was attractive because it would enable the Bear River Migratory Bird Refuge to convert its unused water rights to storage rights to supplement its historical use of 280,000 acre-feet of river water.

None of the sites was perfect. Each had its own set of problems. Avon and Mill Creek wouldn't provide much storage capacity, and Mill Creek would destroy wetlands that would be difficult to mitigate. Oneida Narrows would eliminate another 3.3 miles of

A dam at Honeyville would cover thirteen miles of the river and extend almost to the Cutler Dam tailrace. Peterson Park near Deweyville would be lost.

prime riparian habitat, require replacing PacifiCorp's hydropower plant, and inundate campgrounds that PacifiCorp was required to maintain as part of its licensing agreement. Sedimentation would be a problem, but a reconnaissance report dismissed those three miles as a possible fishery. Honeyville and the Barrens area close to Amalga in Cache County seemed to be better alternatives—if you left out the human equation.

The Barrens option was an off-stream reservoir that required pumping to move the water into the basinlike depression. It would create an obtrusive dike system, and a major power line and industrial-sewage lagoon would have to be moved. Most egregiously, it would inundate a difficult-to-replace fourteen hundred acres of wetland and mudflat habitat and wipe out waterfowl hunting and wildlife observation. The Bridgerland Audubon Society in Cache County was furious and went on the offensive.

Honeyville seemed to be a relatively good choice. Although the reservoir meant the loss of thirteen miles of river, extending almost to the Cutler Dam tailrace, that section was only a Class 3 fishery. There would be some wetlands loss, although not as much as at the Barrens and more easily mitigated. Portions of two state highways would have to be replaced. The springs that provide culinary water for Garland and Tremonton, Utah, would be inundated. Two parks would have to be replaced. The state acknowledged that the reservoir would take out the Hampton stage stop in Collinston, which is on the National Historic Register, and several homes. Otherwise the site provided more storage capacity than any of the others, approximately 117,000 acre-feet, offered new boating opportunities, and had "insignificant visual impact,", according to the 1992 Utah State Water Plan. It seemed to be the leading choice. But not everyone in Box Elder County agreed.

Any of the dam scenarios required Willard Bay as a holding pond for later water treatment and conveyance to Davis and Salt Lake Counties. A pipe and pumping operation would move water from the Honeyville Reservoir into Willard Bay. To accommodate the inflow, Willard Bay dikes would need to be raised, especially since they had settled over the years. In 2001 Representative Jim Hansen, R-Utah, got Congress to appropriate two million dollars to study the environmental implications of raising the dikes. The state proposed enlarging Hyrum Reservoir and putting a dam on the Bear at Honeyville, with the Barrens as a secondary possibility. Momentum was building.

Bear River Migratory Bird Refuge:

Braided Water and Politics at the River's Delta

Hot summer dust devils suck funnels of choking dust hundreds of feet into the air. What were once teeming-with-life wetlands are bone dry, the silty bottoms cracked deeply, like fissures on parched lips. It's not what you expect to see at a waterfowl refuge that must play host to millions of migrating birds in just a few months.

But that's the scene in September at the Bear River Migratory Bird Refuge on the delta where the river empties into Great Salt Lake. During most years, three out of four acres on the refuge are dried up by then, and the Bear River appears geriatric as it enters the refuge. Even a salt-caked, revolving office chair left at a public fishing ground outside the refuge seems to have more kinetic potential than the river.

Nevertheless, hungry birds are on their way, expecting to dine at one of the most important refueling stops in the Pacific Flyway. More than two hundred species of migratory birds will linger and fatten on the marshes. In fall, up to half a million ducks and geese congregate on the refuge's available water. The tundra swan flock swells to more than fifteen thousand. John C. Fremont was so overwhelmed by the concentration of birds that in 1843 he wrote, "The waterfowl made a noise like thunder."

Refuge manager Al Trout, a person not prone to overstatement, explains the refuge's appeal: "Because it's located midway up the continent in a desert, it is a strategic location. It's like a gas station for the birds, an oasis. You've got to come a long ways to get here, so the size and quality of these marshes and location really fit together to be a wonderful location and very important."

"How important? What would happen if the refuge weren't here, if it weren't able to provide that habitat?" I ask.

"We provide habitat to millions of birds," Trout replies, emphasizing millions, "both migrationally and as a breeding area."

But the refuge is threatened by water development on the Bear. An alphabet soup of different rights are stacked at the front of the line and claim water before birds. Trout lists the baffling array: foundational rights, irrigation rights, storage rights, seasonal rights, senior rights, junior rights, flow rights, quantity rights,

The Bear River Migratory Bird Refuge at the delta of the Bear River is a critical feeding and resting stop for birds in the Pacific Flyway. It plays host to more than two hundred kinds of migrating birds every season, as well as the adaptive American white pelican, which feeds on the refuge's abundant populations of carp.

natural-flow rights, diligence water, generation rights, and finally, stock water rights.

Right now the refuge has access to the average 1.2 million acre-feet of water that pass through the Corinne gauging station on their way to Great Salt Lake. The problem is that the flow isn't even, and that variation poses great challenges for Trout and his refuge managers.

"When the refuge was established in 1928," Trout explains, "we got a certificated right from the state engineer for 1,000 cubic feet per second. That was a year-around right. It wasn't like an irrigation right that's got a seasonal basis. That's really the backbone of our water rights. It's a good water right, but it's not the best. We see the 1,000 feet as our foundational right."

"Are you able to sustain that throughout the year?" I ask.

Continued on next page

The yellow-headed blackbird finds the freshwater marshes and reedy ponds of Bear River Migratory Bird Refuge ideal habitat.

"No," Trout replies. "The way that comes out is that we had to prove up use on that in order to file [the right] as quantified. The Bear River never has been adjudicated, and that's in the process right now. You might recall in the early nineties, there was some amount of controversy over a water filing we made, and that was because the state engineer called for all water claims on the Bear to be submitted so that they could publish those, put them on record, and then go through the adjudication process, and then take a right like our 1,000 second-feet flow right and also put a quantity on that."

There is a difference between a flow right and a quantity right?

"The way this works is that the 1,000 second-foot right says that we can divert up to 1,000 second-feet of water every day of the year. Now, the truth of the matter is that, in reality, the river doesn't flow at 1,000 second-feet every day of the year. If it would, we'd be taking 1,000, and it would be that simple. But what happens is in the spring, it might flow at 3,000 or 4,000 second-feet, and in midsummer, it might only flow at 100 second-feet. So we can get credit up to 1,000 for every day that we show we diverted 1,000 or more. Then, on the days when it wasn't there, we can only claim what we've been diverting. So it comes to less than that."

"If you take, for example, the 1,000 second-feet right," Trout continues, "that's about 2,000 acre-feet the way it's converted. One second-foot is 2 acre-feet per day. You figure you'd come up with about 760,000 acre-feet per year if you diverted 1,000 second-feet a day. Ours is much less than that. We'll be somewhat in excess of 400,000 acre-feet as our total claim, and that's not just the 1,000 second-feet. That's adding in some diligence water that we've been able to get. We've purchased some other water rights, some other lands that have water rights that have been utilized before 1902. So, according to the law, if you can prove use before 1902, you are just grandfathered in."

Running a migratory bird refuge is as much about moving water from place to place as anything else. Winter flows are fairly stable with, ironically, clear water. But Trout lets that water bypass the refuge and flow into the lake since

The refuge supports a large population of mammals like the skunk, which stalks the cattails and bulrushes for amphibians, birds, or other small mammals.

the refuge is iced over. He only diverts a bit to sustain muskrat and invertebrate populations.

In April and May, some of the snows begin to melt from higher elevations, and the river flow increases dramatically. Filled with silt, the water loses its clarity, and Trout avoids running that water onto the marshes. Instead, he uses the bypass canals to move most of it directly into Great Salt Lake, to avoid too much sedimentation of the marsh.

"We've done that for years," Trout explains, "and it [sediment flooding] has not had a good effect on the marshes. It tends to raise the bottoms up; also the natural undulations tend to get filled in—the alluvial channels and so forth. We want to keep that to a minimum. So, in the spring, the name of the game is to maintain pool levels without going over too deep. And bypass everything we can on the Bear through our impoundments and into the other areas."

August is the greatest challenge because that's the time when the river flow is lowest. It's also when botulism poses its maximum threat to the refuge. In fact, the deadly disease is one of the reasons why the Bear River Migratory Bird Refuge was created. After settlers moved into the Bear River Basin and began to divert thousands of acre-feet of water, the marshes at the delta began to dry up. By 1920 only two or three thousand of the original forty-five thousand acres of marshland were left. The low water made conditions ripe for botulism outbreaks. More than two million birds died in a 1910 outbreak, and a million and a half in 1920. Congress responded to public reaction to the deaths and in 1928 passed an act establishing the marshland as a national wildlife refuge.

"The refuge has a big botulism problem because at the end of the summer, like in August and September, that's when everyone wants the water," according to biologist Alice Lindahl. Senior rights holders, like the Bear River Canal Company, are first in line and time. It's difficult, then, for the refuge to flush its lagoons with fresh water during low-water periods when upstream irrigators lay claim to it.

In a 1994 study on the water requirements for marshes of the Bear River delta, John Kadlec and Stephen Adair noted that botulism management during summer is limited.

Continued on next page

Their analysis suggested that the refuge only has enough water for low-level management, defined as draining 50 percent of permanent acreage and flushing the remaining 50 percent, only one out of every two years. "Because severe botulism outbreaks occur sporadically and the exact environmental conditions causing the outbreaks are unclear, there is only about a 50 percent chance that water will be available to control waterfowl and shorebird losses to botulism poisoning in the Bear River Delta," according to Kadlec and Adair.

Trout is more circumspect in his appraisal, recognizing that scientists still have questions about the pathology of botulism. "Some folks have said that the lack of fresh water flow and stagnant marshes are what breeds botulism, and that's not exactly true," he explains. "We're finding it more in the marshes that seem to be more midgrade, not so rancid."

"But botulism is water borne?" I ask.

"I don't know if we have to say it's water borne or not," Trout replies. "I think it will just live in mud. But it has to be associated with some degree of wetness."

Water management is a tool Trout needs to control botulism. "A lot of what we do in botulism management is affecting the bird, the host, as much as it is the organism," he explains. "So you have both of those issues in mind when you are playing with water during a botulism die-off."

Trout points out that birds are stressed during that time. "And then some of the decisions you make might be in accordance with drawing birds off an area where they're dying, refilling another area so you can draw birds off of that [the infected area]. It may not have anything to do with reducing the number of microbes in the bad area, but if the birds go off of it, then you've separated the two. So there's some of that strategy, too."

But the refuge needs water to fill empty ponds to make them attractive to disease-vulnerable birds, and that's hard to find in late summer. Trout has a few diligence water claims, but they don't amount to much. He has shut down junior irrigators upstream under severe drought conditions, yet even that water hasn't given him all he needs.

Honeyville Controversy

When Utah water developers started the conversation about building dams on the Bear, the first ally they sought was the Bear River Migratory Bird Refuge. They touted a dam at Honeyville as a way for the refuge to convert a flow right to a storage right. Storing it at Honeyville would enable the refuge to get more water in late summer. But Trout rues the day that he entered that conversation with the state because it has left a legacy of mistrust regarding the refuge's intentions in diverting Bear River water.

"When we applied for our water rights in the early nineties," Trout explains, "that created some concern for folks in the state Division of Water Resources. They were saying that that would be a problem for us to try to get that water in the system the way it is because there is no sum of water. So they came to us with a proposal and said, 'We're looking at the Honeyville site. We want to develop that. But if you would enter into it as a partner with us, we could go in as halfers and split half the cost and split half the water."

"So we said, 'That's an interesting proposition,'" Trout continues. "It might be something that would work for us under some conditions. Number one, that we fully mitigate our wildlife impacts. Just because we're partners doesn't mean we erase the need for mitigation. Number two, we would pay for half the water and get half the water, pay for half the cost and get half the water. And then we were also looking at protecting the rest of the corridor with some possible conservation easements from the refuge to the dam, creating some minimum flows."

"But what happened on Honeyville," Trout

By the end of the summer, three out of four acres in the refuge dry up. If more water is diverted from the Bear River for municipal use along the Wasatch Front, refuge caretakers will have fewer management options and more difficulty trying to protect a threatened ecosystem that relies on water from the Bear.

explains, "was that we did some initial analysis. We had to convince U.S. Fish and Wildlife people on this. My first hurdle was to get our Utah environmental ecological services folks on board. After we looked at it and talked about what could be done, they got on board with it. They said, 'Yeah, this looks like it could be doable.' So we actually went to Washington with the proposal, took all of our numbers and all of our charts. We made the presentation to the national ecological services folks, and they bought into the fact that we should go ahead and continue on with the process. Now we did not get a final approval to sell it. What we did get was a green light to proceed to the next round of planning. We came back with that news, all happy—and the state got cold feet at that point."

During the interim, local water officials began to rethink whether they wanted the refuge's participation. There was a downside they hadn't contemplated. "Some of the locals, as well as some other folks," Trout explains, "started looking at it and then said, 'Hold it. You know, Honeyville is the cheapest water to be developed on the Bear.' At that time, they thought it was their best dam site. And they said, 'You're going to give the refuge half right off the kick start, and that's going to yield enough [for them] completely. After Honeyville's built and [we want] to build more dams,' they perceived us as not cooperating

Continued on next page

During heavy runoff periods in late spring and early summer, when the Bear carries a high sediment load, managers direct the water into conveyances like the Reeder Overflow Canal so that it bypasses the refuge and flows onto the Great Salt Lake mudflats. Scientists have found that the excess sediment raises the level of refuge ponds and diminishes habitat.

because we would have been taken care of. They perceived that the refuge wanting water would aid them in the development of all the dams they wanted to put on the river."

I laugh at the state's exercise of raw politics. "So, as long as they put you last in line..." I suggest.

"Exactly. And they were honest, at least, about that," Trout replies. "I thought it was entirely an unfounded argument. I just didn't see the logic in it. I don't think they understood how these environmental decisions are made within the Fish and Wildlife Service. They felt like if I just raise my hand and say to my folks, 'Well, hey, look, we want some of that water,' then they would just approve any dam as it came along with a rubber stamp."

"So we withdrew," Trout continues, "and over the years, there has been a lot of confusion. They jerked the rug out from underneath us, basically, and said, 'We're going to build this alone, and we don't want you as partners.' At first, they saw us as a partner that would help them. As opposition generated, they saw us as a liability, not an asset. At that point, we've no longer been associated with any Honeyville proposals. And it's been really hard for me to get that word across to people because when the word 'Honeyville' comes up, a lot of folks will say, 'Oh, that's the one the refuge is involved in.' No!" Trout emphasizes the word, the first time he's done that during the interview.

Alice Lindahl echoes Trout's feelings regarding Utah's Division of Water Resources. "The refuge has a historical right that actually precedes a lot of the irrigators," she says. "So the state is sort of covetous of those, and they also see the feds as having really deep pockets. 'If we can get these guys to pay for half the dam, we're halfway there, and then we'll stick it to them later.' That's the kind of impression you get from hearing them talk. They just are using the refuge as a money source. They don't care at all if the birds are in better shape or not better shape."

Instead of Honeyville, the refuge is promoting Hyrum Dam on the Little Bear River as a storage site for its unused water rights. It would be able to store spring runoff there and use it later in the summer. More importantly, that would enable the refuge to develop its acreage fully and be more in control of its water, rather than relying on the state's goodwill.

To deal with a diminished water supply in late summer, the refuge is beginning to divide its units into smaller subunits. That enables managers to direct limited water to the most critical habitat more effectively.

Under this fifty-million-dollar proposal, the Hyrum Dam would be raised approximately fifty feet, allowing the storage of an extra 40,000 to 50,000 acre-feet of water. That would allow the refuge to convert its 1,000 cubic feet per second junior right to a storage right. The larger reservoir might help alleviate algal blooms at Hyrum, but it would inundate the state park. Park officials aren't necessarily against the idea, even though it will require moving the campground, because it will allow them to be farther away from the town. So Trout feels it would be a good project, and he can justify the refuge's water needs based on sound science. Right now the refuge wetlands shrink to about ten thousand acres in August. By developing its full water rights, the refuge would only dry up five thousand of its total acres.

Until then Trout has to take intermediate measures to husband what little water the refuge receives in late summer. Instead of filling only Unit One and letting the other four become sun baked, the refuge is constructing new dikes to create a system of smaller ponds.

"We can just be much more efficient with water use with smaller units," Trout explains. "The original ones were five thousand acres each. We want to come up with about thirty units now. We've divided all of our units now, except for one, into smaller subunits. We just identify which units we want to keep. We'll divert the water we have into those. We don't want to keep too many wet and lose them all late in the summer anyway."

Trout is nervous about any diversions of the Bear River but says that they likely will happen.

Continued on next page

"It's just a matter of time, whether you're talking twenty years or fifty years or a hundred years," Trout says. Right now the refuge has the use of that undeveloped water before it empties into Great Salt Lake. If it's diverted to the Wasatch Front, and if the refuge can't find a place to store spring water, that will create an even more precarious situation.

"We've got 1.2 million-acre feet a year now that we're getting," Trout sums up. "We have a lot of room for error because we've got a lot of extra water, and we can make up for it. But if we don't go out and get boards changed—and basically we're not staffed like we need to be for that intensive management—we can goof up. We've got everyone else's water right now that hasn't developed it for cushion. There are times of the year we don't have, but right now we've got more than our 1,000 coming through."

If the Bear is diverted, there will be less margin of error for the refuge, and that will put Trout in a tenuous situation. He is charged with protecting millions of birds that are part of the Bear River community but have no voice. He senses their increasing vulnerability as he asks people to balance the needs of waterfowl on the scales of justice that are heavily weighted on the side of human water politics.

Political Pressure Hits the Utah State Legislature

Momentum was building on another front, too. A group of concerned Box Elder ranchers scraped together the resources to produce a video entitled "We Are Concerned" that outlined the negatives of the Honeyville Dam. The URC took its argument to the media, and eventually to the Utah State Legislature. The URC pointed out that more than just a few homes would need to be replaced. More than fifty farming families would be displaced, in addition to the impact on water, sewer, and irrigation drainage systems. The reservoir would take out valuable farmland, a special concern considering how important farm receipts in Box Elder County are to the state's breadbasket. The URC argued that Honeyville Reservoir would be almost drained by the end of the summer season, becoming a shallow, barren mudflat unusable for recreation, only good habitat for cockleburs. It also pointed out that the dam site overlaid the Wasatch Fault and would be vulnerable during an earthquake since the silt and clay lands surrounding it would liquefy.

The URC joined forces with the Northwestern Band of Shoshones in arguing protection for the archaeological sites and buried tribal remains along the river, although periodic flooding had unearthed Indian bones in the past. The Bear River Migratory Bird Refuge signed on as opposition, too. Not trusting that the Hyrum Dam would be raised, refuge management said that it would lose 20 percent of its water at a critical time when already three out of four acres of wetlands dry up each summer due to upstream agricultural use of water, thus putting 220 migratory bird species at risk at a vulnerable time in their yearly life cycles. Then the URC teamed with Cache County birding interests to mount similar arguments about critical habitat loss if the Barrens dam was built.

The URC proposed a set of alternatives that would be far cheaper than putting dams on the Bear as well as more environmentally and socially sensitive. First, it suggested water conservation for the second-driest state in the nation, which ironically has one of the highest per-capita water-consumption rates. A 20 percent reduction in water use would free up 44,000 acre-feet of water at a cost of pennies per acre-foot. Second, it took advantage of 1992 adjustments to the Central Utah Project (CUP), which moves water from the Colorado River Basin west into the Great Basin, to make the heretical argument that it was silly to send CUP water to Juab County to provide agricultural water to fifty or sixty farmers to grow alfalfa. The cost for moving those 30,000 acre-feet of water north would be less than $100 million, and besides, that water was likely to be converted to high-cost municipal and industrial water anyway, creating a windfall profit for the Juab farmers. Third, the URC suggested accelerating JVWCD's conversion of surplus irrigation water in Salt Lake County to culinary use. Because of urbanization, the water was no longer used for irrigation and ran through backyards and subdivisions more as ornamental landscaping. That source would amount to approximately 60,000 acre-feet. Moreover, some ditch owners had reported that they had offered to sell their irrigation water to the JVWCD but hadn't gotten a nibble.

Opposition arose from an unexpected quarter, too. While the Bear River near Honeyville meets the definition for recreational and wildlife uses, it doesn't meet the criteria for safe drinking water. Costs for treating the water for total dissolved solids, turbidity, hardness, iron, and manganese would be substantial, and that made Weber Basin Water Conservancy District uneasy. Since it controlled the water in Willard Bay, it did not want to see its quality degraded by water pumped from the Bear River. Also it had never been clear to anyone how the water in Willard Bay should be used. Originally it was supposed to supply water for agriculture. Later, it was rediscovered as a warm-water fishery and boating destination. Then a mineral facility made claim on some of the water.

Environmentalists seized upon the lower quality of Bear River water and confusion about Willard Bay. They welcomed Hansen's appropriation because they believed that it would show that a large share of water in Willard Bay wasn't used for agriculture, possibly freeing up still another supply for the burgeoning Wasatch Front that was better quality than what was in the lower Bear at certain times of the year.

After public response, cost analysis, and a revised water-needs estimate, the Utah Division of Water Resources modified its plan. It still called for using Willard Bay, a connecting pipeline from the Bear River at Elwood to Willard Bay, and conveyance and treatment facilities to move the water to the Wasatch Front. It altered the fourth proposal from building "a dam at Honeyville" to "a dam in the Bear River Basin." The revised plan seemed to abandon the Honeyville and Barrens dam sites, a canny move by division director Larry Anderson. But the division warned that, because of the extended period of time the plan covered, it could be modified again. The stage was set for a fight in the Utah State Legislature.

In 2002 the URC prevailed on Senator Michael Waddoups and Representative Eli Anderson to submit a bill officially taking the proposed Honeyville and Barrens dams off the table. Other dam sites could still be considered, with attention starting to focus on the Washakie site near Portage, Utah. The legislature passed Senate Bill 92, and Governor Mike Leavitt signed it into law, partly because it was largely symbolic since the division had earlier retracted the Honeyville and Barrens dams due to public opposition. But Senate Bill 92 did seem to be the final nail in the coffin.

Then the URC received some other good news. When the Juab County recipients of CUP water announced plans to resell it to urban users after the project was completed, it freed CUP management from their pledge to supply agricultural water. That made the URC proposal to send it north more politically feasible. In 2002 the CUP announced that the Spanish Fork/Nephi irrigation system would be scrapped, partly because it had been challenged by the Strawberry Water Users Association during the scoping process, and that water would be sent north. The CUP would help Juab County farmers tap groundwater for irrigation instead.

That momentarily eased the threat to the Bear River. It had been a good year for the URC, the Box Elder farmers, and the Northwestern Band of the Shoshones. But Wasatch Front population growth curves still sloped inexorably upward.

Yeoman Ranchers Challenge the Dam Builders

Members of the Shoshone tribe were baptized in the Bear after they converted to Mormonism soon after the Bear River Massacre. By 1876 an area between the Bear and the lower Malad River was called Indiantown, where Mormon missionaries helped tribal members make homestead claims. At one point, there were close to fifty homesteads. But due to the problems of trying to use the Bear to irrigate the sandy soil, most of the Indian homesteaders only stayed five years. Many of them moved to the Washakie settlement near Portage that George Washington Hill had obtained for the Northwestern Band of the Shoshones.

"It doesn't sound like a lot," Laura Selman says, "but you've got to remember that the Indians that homesteaded here were the first Native Americans that homesteaded anywhere in the United States. They gave up their tribal ways, joined the [Mormon] Church, homesteaded the ground, and began to be farmers."

At one spot near the river, missionaries showed Indian homesteaders how to use gravity to allow two springs on a bluff overlooking the Bear gingerly to nourish their small crops of melons and potatoes. Laura and

Fred Selman works the land on the family ranch overlooking the Bear near Deweyville, Utah. The river provides the Selmans a living and is an integral part of their family history.

Fred Selman have a special fondness for the Shoshones, partly because they share a common history with these springs and the land next to the Bear near Deweyville, Utah. Their 350 acres include several of the old Indian homesteads.

Born in 1941, Fred grew up on the ranch his father purchased in 1944 and bought out his brother's interest in 1996. The ranch has been his life partner, or had been until he convinced Laura, daughter of local dry farmers, to share the well-watered land with him. Together they envisioned doing something else with the two springs. They diverted them into a canal and piped the water to a fishpond they dug out below the bluff on their river bottomlands. They also use the water to irrigate an acre of grass around the pond.

"Is it kind of like your family park?" I ask.

"Yeah, that's exactly what it is," Fred replies. "The Indian homestead used to be called Lemuel's Garden. So we've named it Lemuel's Garden Park." The local Mormon ward uses the Selman oasis for many church activities.

The Selmans stock the pond with trout, although only brown trout do very well in the slightly warmer, fifty-eight-degree water. Fred figures they'll have to drain the pond, though, because carp somehow got into it, probably from a passing pelican that couldn't carry a

Continued on next page

full load. They don't plant the trout as food but simply to make the family park more enjoyable for the kids by offering fishing.

"The kids—that's what it's all about. You know, I've got to say one thing," Fred explains in the longest exchange in our interview. "We say we raise so many cattle and we raise so many sheep, and the horses and everything else, but you know, it all boils down to the fact that this is a good place to raise kids. That's what it's all about—whether it's our kids or the ward's."

Laura and Fred are old enough to have witnessed most of what the Bear River has to offer. During the drought years of the early twenty-first century, they saw their production drop by 25 percent, even though they switched from corn to barley since it takes water earlier in the growing season, when there's a better guarantee that the Selmans will have it. Most of their grain goes to feed their two hundred head of cattle and twenty-five hundred sheep, with a bit left over for sales. But in 2002, they had to buy hay for their sheep, which were trying to eat the meager pickings on the their BLM allotment in the Hogup Mountains. In spite of the drought, though, they've still found enough water to irrigate fruit trees and a vegetable garden, where, in summer and fall, people come onto the ranch and help themselves to whatever's freshly picked, stacked, and modestly priced and leave money under the honor system.

The Selman ranch has seen high water, too. Although Cutler Dam approximately ten miles upstream from their ranch minimizes flooding, Cutler overflowed in 1983 and flooded their bottomlands with six to eight feet of water. They lost a few acres to a new channel that the river cut through the rich farmland, but rather than trying to reclaim it, they turned it into a riparian corridor, letting the willows grow in from the river.

"To be truthful to you," Fred admits in a kind of confessional voice, "we decided we'd turn it back to wildlife. It's good bird habitat and big game. Rather than making a big fuss [to government, asking for flood-relief funds] and spending a lot of money—we'd talked about taking the dirt on the other side and leveling the field—but after we thought about it, we decided to just turn it back. It's a jungle down there now."

The ranch didn't get a lot of new, revitalizing soil from the flooding, either. One of the downsides of dams like Cutler is that they prevent the natural regeneration of soils by holding back sediment that normally would be spread out during flood times.

When UP&L had to release water from Cutler in April 2005 due to heavy runoff, 8,000 cubic feet per second came rushing by the Selman ranch. Their pastures and cropland next to the river were flooded, and they figured they'd have to replant the grass around the pond, although they knew the pond itself would be safe since Fred had moved it up higher from the river after the flooding of 1983. But, all told, the Selmans didn't lose too much in the recent flood.

"We knew it was coming," Laura says. "When you live on the river, you expect it. Yes, it hurt us. But we don't blame anyone."

The Selmans' rapprochement with the Bear is born of necessity. Its water courses through the family genealogy, just as its floodplain provides them a living. They deal with the daily four-foot elevation fluctuations from Cutler. They pump 1.5 cubic feet per second from the river to sprinkle their adjacent land. They use water from the West Side Canal of the Bear River Canal Company to irrigate the rest of their acreage.

They hunt on the river. The neighborhood kids play in their garden park next to it. But there is one rule all kids must obey: they can't get close to the river. If they balk, Laura reminds them that Fred lost a grandfather and a friend when they drowned in the Bear.

In a few slow moments, usually during winter, Fred finds time to walk the river. It's

Laura Selman cares for the lawn at their well-kept home. She and Fred teamed with other farmers and ranchers opposed to the dam site at Honeyville and produced a video entitled "We Are Concerned."

then that he occasionally discovers historical artifacts that have found their way into the Bear during the river's short (in geomorphic time) interlude with people. On the old Indian homestead, Fred found a bullet mold for a Browning rifle dated 1840 in Quincy, Illinois. He figures it belonged to a trapper. He picked up a Henry rifle near Boise Ford, and a cast-iron harrow tooth for turning the ground, probably dating from 1870, when the land in the area was first farmed. The river unearths ox and mule shoes every once in a while, and Fred has stumbled across some finger bones, too, but no dead bodies. He does note that they've found a skull and other remains of wooly mammoths along the river.

While the family has a long attachment to the river, they never supposed their relationship would become political, too. But when the state proposed building a dam at Honeyville, it was a direct threat to the Selmans' way of life. They became activists and, with a group of local farmers and ranchers, produced a video titled "We Are Concerned" that documented what a dam at Honeyville would destroy. The reservoir would inundate about a hundred acres of their rich bottomland. Water from the reservoir would wreak havoc on septic and field drainage systems and cause erosion.

"It was such a big threat to our canal systems for the rest of the valley," Fred says, because it would have backed up some of the groundwater. "It would have put those cutbanks in jeopardy [for erosion]. It would have shut down the natural drains of these farms—not only the natural drainage but the drainage systems in these farms—because it would have put the water table higher than these drainage systems. But we couldn't make them believe this."

"And not only that," Laura continues, "this is sand. You're not just going to flood these homes; you're going to keep on going [the spreading groundwater will displace more families]."

Laura then describes the public perceptions to their opposition to the Honeyville Dam. As ranchers, she and Fred felt uncomfortable being perceived as against water development, a commodity they value as much as others do because it's the key to their livelihood and future.

Continued on next page

"People get the idea that we were totally against the development of the water on the river. That wasn't what we were against. What we were against was that they weren't developing a quality product. What it was doing was taking the last green space that's available in Utah and turning it into a swampy marsh," she explains.

"A drain-and-fill project," Fred adds, describing the land above the dam as a mudflat filled with nothing but cockleburs most of the season.

"It wasn't a recreation site. What are you leaving your posterity when you can't do better than that?" Laura asks rhetorically. "If you're going to take water and develop water, develop a quality product."

The Selmans would lose their Indian homestead springs. Tremonton's springs, from which the city pumps culinary water, would be inundated, too, taking away that source of water for Garland and Elwood.

"With no way of knowing where replacement water would have come from?" I ask.

"Well, they told us they would dig down and catch these aquifers before they come into the river," Fred replies. "And I argued, 'If you can do that, how can you trap the water from going the other way? From the reservoir following the aquifer back?' So they could never give us a straight answer on that. That's the kind of logic we were up against. Or no logic."

Not every farmer in Bear River Valley agreed with the Selmans. Charles Holmgren felt the Honeyville Dam would be a good thing for the Bear River Canal Company. It would allow members to "move" water rights more easily between the East Side and West Side Canals.

"The Honeyville dam is very controversial, locally," Holmgren says. "I know people would be impacted. Maybe I would have a different tune if my property was near the Honeyville dam. But the Honeyville dam is about the only one that would be affordable to benefit the canal company."

"I look at a Honeyville reservoir, and, in my mind," Holmgren continues, "I would imagine something similar to the Hyrum Reservoir or the Pineview Reservoir. For most of those people, if you were looking at it from a financial point of view, I would think it would increase the value of property. Of course, farmers would swear they're not in it for development, but, ultimately, that's what you end up doing with your farm—selling it for development. I think it could be an asset to the valley, but you get Fred and Laura Selman who are just adamantly opposed. And the mayor of Elwood is adamantly opposed to it. And you go down, and a number of people are opposed. I don't want to sacrifice them, but I look at the rest of the valley as being sacrificed because of their philosophy to it."

Laura and Fred Selman have their eyes on the future, even if they are in their twilight years of farming. They are pleased that their son wants to continue the family business. But they are concerned about the long-term outlook for farming. That subject comes up when I ask if they like to take twilight walks on the river.

Laura immediately laughs. "We don't have much time to go for twilight walks. The business we're in, people say, 'Oh, wouldn't it be wonderful to live here and have this grand life.' They don't realize the work that's involved in it. I think he [Fred] started at a quarter to five this morning, and he's been at it all day. That's our way of life."

"People don't realize what's happened to agriculture in the last ten years," Laura continues, her voice reflecting a sense of concern. "People don't realize how close they are to being hungry. Your grain markets are down. Your dairy has had problems. Your cattle markets are not stable. People don't realize where their food comes from.'Why do they need the farmer?'" she asks with tongue in cheek. "'I can buy anything I want at the grocery store.'"

Laura then expresses the frustration that many farmers and ranchers have with some

The Selmans enjoy a rare moment of leisure, fishing in the family park they call Lemuel's Garden Park. They dug the pond, then filled it from a spring once used by Shoshone homesteaders. The Selmans have entered an agreement in principle with the Nature Conservancy to forego selling their ranch for real-estate development in return for a conservation easement.

factions of the environmental movement. "They want to buy off your BLM permits. They want to buy off your forest permits. They want to turn it all back to wilderness. And we're going to ship our food in."

"Do you realize," Laura continues, "that if you start shipping your food in and you start doing away with your agriculture, what's going to happen to your state? We could become a third-world country very easily if we don't maintain our agriculture. And the farmers are tired. There's big prices out there. So they're selling that ground for big prices. But you only build one crop of blacktop."

Fred turns to Laura and asks, "The average life of a farmer now is what?"

"The average life of a farmer now is sixty years old," Laura replies. "And there's no one going into it. I've worked over at the Farm Service Agency for twenty-five years, and you used to see these young people come in with their dads. Now it's the guys Fred's age that are coming in, and they're tired and worn. They're tired of paying high prices for their machinery, and yet we see low prices for the commodities being produced."

"I think one of the big impacts that works against the agricultural dream is the strength of the American dollar," Fred continues. "The other countries are enjoying the marketing of their commodities here because the dollar is worth twice as much as their's [country's currency]. So that makes it so very attractive to them."

Over the past ten years, the Selmans have noticed that their operation is increasingly affected by the global economy. Like all farmers, they face downward global pricing pressure, while supply costs accelerate. "Imports are killing us," Laura says. "They're shipping barges of wheat in here; McDonald's is buying their cattle from Australia; the entire sheep market is over buying in New Zealand."

Still, Laura and Fred Selman are content and have a rich life, even if their tidy house isn't that large. They only need a few things: clear springs, water from the Bear, good land for raising crops—and the next generation.

Future Water Development

When David Ovard, general manager of the JVWCD, looks at the charts, he gets panicky. The population forecasters say there will be an additional 1.5 million people along the Wasatch Front by midcentury, and probably sooner. They'll require 500,000 acre-feet of water, water that he and other wholesalers don't have. When you consider the Bear River Development Act only promises his agency 50,000 acre-feet, the numbers paint a doomsday scenario.

"The cheap water, the easy projects, they're all done," Ovard says. "Everything we do is going to be more expensive. If we could reduce water consumption in the state by 25 percent over the next twenty years, we could save a billion dollars in infrastructure costs. That's a billion dollars less in tax money, in rates, in impact fees."

"Do you think it's possible in the long run for the state to avoid using that undeveloped water in the Bear River?" I ask.

"I think it's just a matter of how far you want to go out," Ovard replies. "I look at the demand curves. We projected we would need Bear River by 2015. If we are able to do some other things. If Central Utah is able to make a little more water available to us. If we can implement some conservation. In my mind, all we're doing is pushing the Bear River project back."

"A million and a half people, 500,000 acre feet of water," Ovard continues. "Do you realize how much water is available to be developed in the state of Utah period? Eighty percent of it is in the agricultural community right now. And I know that they're working on trying to be more efficient. If you have the Bear River with [approximately] 200,000 acre feet, and other than that a little dab of water here and there. And groundwater is largely gone throughout the state. The local water systems are all tapped out. Reservoirs are very controversial, even if there were some water rights, which I'm not aware of any. So the Bear River is probably the most economical place to get a large enough quantity of additional water other than conservation and wastewater reuse."

Ovard is keenly aware of his particular place and time and role in Utah's development. He knows what it takes to bring water to homes, unlike most citizens, who don't give it much thought. All they have to do is turn on the kitchen tap. Water is always there.

"The history of water development is that farsighted people have done the things that are necessary for future generations to make sure they will not be without water," Ovard says. "That's all I have in my mind. The water people have been out ahead of development generally. Some of my greatest fears are that some things the governor is doing now are very shortsighted."

Ovard is referring to Governor Mike Leavitt's proposal in 2003 to redirect the state's dedicated one-sixteenth of sales tax from the water-projects development fund to education. He wanted to recapitalize all loans and use that money for public schools. Leavitt was looking at another frightening chart: the projected growth of student enrollment and the already-underfunded education budget. He was trying to balance the scales weighing Utah's two Holy Grails—education and water. Ovard feels that was shortsighted and might cripple water development for future Utahns.

Ovard does have one ally in his fight to maintain, if not expand, money for water development—drought. It has a way of bringing issues to the foreground. And with 1998 to 2004 marked by drought, six years of conservation messages were helping people to see the big picture the way Ovard does.

"Maybe it's going to take that extended drought and be brought to our knees," Ovard says. "Right now the state is really taxed as far as meeting water needs. We may get through this cycle. We may go right through it and somehow come through it with no change. Then maybe it's the next one. If this one lasts long enough, we go a couple of more years.... It's going to take some water systems to crash. It's going to take people to be out of water, and it's going to require probably being a state and national disaster, and we'll have to get federal aid. And everybody will say, 'This is a terrible problem' before we really get serious."

Our conversation then turns to the more immediate problem of trying to develop Bear River water if dams are off the table. Moreover, if the only storage facility is going to be Willard Bay, that water still has to be moved to Davis and Salt Lake Counties. How will that be done?

Ovard is happy that more CUP water will be available for Salt Lake County now that the Juab irrigation system has been scrapped. But he notes that the water will be more expensive than some people thought, lessening the savings. Once that water is converted to municipal uses, the cost per acre-foot skyrockets because, unlike agricultural water, municipal use doesn't qualify for subsidies from hydropower revenues from the Colorado River system.

Regardless of when Bear River water needs to be developed, part of Ovard's responsibility to be farsighted includes buying land for water-treatment plants and pipeline rights-of-way. The real estate bubble of the early twenty-first century meant that land was getting more expensive, and the JVWCD thought it prudent to purchase some before it became even more expensive. Also the agency only wanted to purchase land from willing sellers, rather than using the heavy-handed club of eminent domain and condemnation. Since 1997, the JVWCD has spent more than ten million dollars for rights-of-way.

His agency also budgeted to purchase land for a water-treatment plant in West Haven City that will be operated in partnership with the Weber Basin Water Conservancy District. Originally the city had opposed the plant siting, but Ovard felt their objections could be met through a scoping process. There was also a comfortable working relationship between Weber Basin and Jordan Valley since Tage Flint had been hired as general manager of Weber Basin, succeeding his father, after working as a staff engineer at Jordan Valley for fourteen years.

Ovard's agency also got entangled with the third Holy Grail in Utah politics—highway development. As part of its lobbying strategy to scuttle dam building on the Bear, the URC claimed that the pipeline moving water south from Willard Bay was being engineered into the controversial Legacy Highway project in Davis County. It claimed that the Utah Department of Transportation (UDOT) had tried to hide the possibility that a sound berm and trail on the west side of the highway could be used for a pipeline to carry Bear River water. Referring to documents obtained through Utah's open-records law, Zach Frankel, then executive director of the URC, said that UDOT had deliberately not used the term "Bear River" in the Legacy Highway's environmental impact statement because it would compound the political problems the highway was already facing with environmental lawsuits.

"It's my understanding that the proposed aqueduct that would bring the Bear River south is tied into the Legacy Highway right-of-way," I tell Ovard. "What would happen, then, if the Legacy Highway were moved, or if it's not even built—what would happen to those plans?"

"It's not tied in at all," Ovard replies quickly, as if he is trying to squelch a rumor. "It's an alternative that we've looked at. You know, Farmington Bay creates such a narrow neck of land between the mountains and the lake it's only a very small corridor to get down through there. We approached UDOT about that segment of the Legacy Highway—possibly putting a future Bear River pipeline in that. We've also talked to UTA [Utah Transit Authority] about putting it in the railroad right-of-way. When it comes time to do the project, if we could lock up, if we could find a publicly acceptable place to come through that narrow neck of land by piggybacking on somebody else's right-of-way, it certainly would make sense. It probably would save some dollars and minimize environmental impacts. But no commitments have been made, and we're still exploring alternatives."

Ovard may have time to explore even more alternatives because in May 2005, Utah's water-financing task force said that developing Bear River water could be postponed. Continuing rapid growth in southern Utah is making the financing of a pipeline bringing Lake Powell water to Washington and Iron Counties a higher priority. Besides, those new water sources from the CUP and the public's good response to conservation pleas during the drought are taking the heat off developing the Bear. The current target date for moving Bear River south is 2025, and the extra breathing space is fine with Ovard, who wants to see if future water needs can be eased with conservation.

In water development, grand vision comes long before mundane execution. "Until someone is ready to sign on the line and start building something, everything is just a concept," Ovard observes. Moreover, in spite of assurances that the Honeyville and Barrens dams won't be built, everything is fluid when it comes to moving water.

"If you are going to develop the full 220,000 acre feet, a lot of it is committed up north," Ovard explains. He notes that was a smart political maneuver by the sponsors of the Bear River Development Act, who wanted to enlist northern Utahns' support delivering water to the Wasatch Front. So the legislators made sure that the act provided assets for Box Elder and Cache Counties. "Whether the development in the next fifty years will need anywhere near that, I have no idea. If you were to develop all of that, you probably are going to need one or two dams in addition to Willard Bay," Ovard says.

The political intricacies fascinate Nancy Mesner, assistant professor in the Department of Aquatic, Watershed, and Earth Resources at Utah State University. She's also an extension specialist who has done research in water quality on the Bear River.

"If I went back and looked at the history of all the dams on the Bear, it's like a deck of cards. They reshuffle the deck, and about four or five come to the top," Mesner says and stops to laugh. "And they become this decade's highest priority. They come and go. I don't think the legislature passing a law makes things go away. As long as there's money and as long as there's...water that is unclaimed or developable, I think it [dams] will continue to be an issue."

Pollution in the Bear River Watershed

Agriculture, including feedlot and grazing practices, is the primary source of pollution in the Bear River Basin. It degrades approximately 312 stream miles, especially those adjacent to the offending operations, about six times as much as resource extraction, which is in second place. Urban runoff is a distant third.

Pollution from pesticides isn't so much a problem in the Bear River Basin, especially in areas, like Cache Valley, which contain more pastureland. Alfalfa and hay don't require a lot of pesticides. There aren't many problems with heavy metals on the Bear, either, because the basin doesn't have marine sediments like selenium, found in shale. Rather, the Bear River watershed has mostly lacustrine sediments from Lake Bonneville.

The two primary pollutants from agriculture are phosphorus and nitrogen, and both are hard to pinpoint and measure. Both can enter the biological system as runoff from fertilizers. Nitrogen can result from discharge from wastewater-treatment plants. Complicating matters, both can be in the system naturally, phosphorus from sediments and nitrogen from atmospheric deposits.

Regulators focus more on phosphorus than nitrogen, though, "probably because it's harder to control nitrogen in the system," according to Nancy Mesner. "So, if you can get control of one or the

Field drainage systems in rural areas often terminate with pipes that funnel return from flood irrigation into the river. The runoff adds to the phosphorus, nitrogen, and pesticide loads in the Bear.

other to prevent algal growth, and if phosphorus is more manageable, then pragmatism says that is what you should go after."

The primary problem with either phosphorus or nitrogen in the aquatic system is algal blooms. Algae feed on these fertilizer nutrients. "What happens when the water slows down long enough for the algae to really start clicking you can start getting into these bloom situations," Mesner explains. "There is a cascading effect, where the algae decays, sinks to the bottom, and sucks oxygen out of a lake as they decay. For instance, Hyrum Reservoir and some of the other smaller reservoirs have bad problems with fish kills—winter fish kills and sometimes even summer fish kills. Usually that's a result of low oxygen, which is a direct result of nutrient level, either phosphorus or nitrogen" since it alters the balance of available oxygen in the system. The result is eutrophication, a condition marked by relatively low oxygen levels.

"There are aesthetic issues and smell," Mesner continues, drawing out the several syllables. "On the lower [Bear River] system where they are talking about taking some of this tract water and treating it—the drinking water along the Wasatch Front—you can get the kind of algae species that cause taste and odor problems," even after the water is treated.

Eutrophication is more a problem in lakes and reservoirs of the Bear River than in its streams. But when the velocity of the river slows, as it does in the lower Bear, the main stem can experience eutrophication. In addition, nutrients from upstream reservoirs can stimulate excessive growth of plants, as in the tailwater below Oneida Narrows Reservoir.

"The sediment has dropped out, and the water is clear, and you often have more dissolved nutrients 'pulsing out' of those reservoirs,"

Mesner explains. "You can get those beautiful tailwater fisheries because you're feeding your system. But what can happen with that oxygen link is that oxygen is being produced by plants all day long, but all night long the plants are no longer producing oxygen, but, just like us, they are using it. So the plants continue to use oxygen all night long, even though there's no production. That can actually drive oxygen concentrations down very low on a daily basis. So higher in the day and lower at night."

Bear Lake is threatened by "cultural eutrophication," according to a 1989 report by the Army Corps of Engineers. Normally oligotrophic (oxygen rich, nutrient poor) Bear Lake has been relatively transparent due to sparse phytoplankton growth. Its unique turquoise color has come from high concentrations of dissolved limestone. But now, because 70 percent of the inflow into Bear Lake comes from the river via PacifiCorp's Rainbow Canal, that water carries nutrients from upstream, farm runoff, which can cause oxygen depletion, especially during spring runoff. With the phosphorus level going up, that imbalance leads to eutrophication. But nitrogen has been going down, somewhat minimizing the effect, although the levels of both phosphorus and nitrogen in the lake are expected to decrease during periods of drought. So the data are inconclusive.

That doesn't ease the concerns of Bear Lake property owners. They always wondered why Bear Lake needed to be treated like a reservoir in spite of the fact that John Wesley Powell decreed it should be the first reservoir storage site in the West in 1889, according to Robert Parson in his *A History of Rich County*. Hydrologist Vincent Lamarra and other scientists proposed more upstream storage to free Bear Lake from any storage at all, thus cleaning the water and protecting endemic species. Recreationists would benefit from a more stable lake level and only face periodic fluctuations due to drought and flooding. But the Bear River Commission disregarded this environmental argument when drawing up the compact. History made irrigation and associated rights on the Bear just too entrenched and venerable.

While regulators measuring improvement in water quality are focusing more attention on quantities of phosphorus and nitrogen in river systems, their results are inconclusive. "I don't know of any data," Mesner says, "that would suggest that things are necessarily getting any worse or better, which is kind of ironic. But the problem is we're looking at a pretty small window of time, considering that the impacts have been ongoing for approximately a hundred years in the watershed."

Mesner adds there has only been monitoring for about twenty years, thanks to the requirements of the Clean Water Act. She feels people need to look at more long-term hydrological cycles as they measure pollutant loads in rivers like the Bear. Moreover, the three states entrusted with cleaning up the river all have different approaches, although the Bear River Water Quality Task Force has

As businesses nestle on riverbanks, storm drains like this one in Evanston, Wyoming, carry nonpoint pollution from urban activities into the river.

made considerable progress in aligning methodologies for assessing impairments.

"Utah is invested heavily in water-column [water-chemistry] monitoring," Mesner says. "So they've got a really good database and will continue to do that. Idaho has done precious little water-quality monitoring."

That apparently comes from both lack of funds and lack of interest. Mesner adds that Idaho "got itself into trouble about nine or ten years ago for not doing an appropriate job of testing all their water bodies—in Idaho, the EPA got sued, as in a lot of other states—and for accepting these water-quality assessment lists. It just wasn't adequate. The judge made them redo almost a thousand assessments. So they developed a protocol that heavily stressed other types of indicators. They'd go in and look at riparian areas and macroinvertebrates, a little bit of fish, and try to go in, take some measurements, get out, and make that their computer assessment. It's a different philosophy and a different approach."

"Wyoming's approach has been to say, 'Bear River? What Bear River?'" Mesner continues, referring to the fact that Wyoming has a lot of bigger, more charismatic rivers. In Wyoming, the Bear River doesn't have as much political power.

"Would you say that the Bear River in Utah has a relatively powerful political constituency?" I ask.

"Yeah, I think so," Mesner replies, "and we're in a very interesting place because of the Bear River Compact, because of the Bear River Basin Water Plan and the legislation [to develop the

Allowing cattle to graze to the banks of the Malad River, a tributary of the Bear, denudes the land and destroys the riparian corridor, increasing the sediment load of the river downstream.

water resources]. So you've got a lot of people gunning for the last water. And that sucking sound is not jobs going to Mexico. It's the Wasatch Front looking for more water."

The 1992 Utah State Water Plan for the Bear River Basin provides an overall assessment of pollution in the watershed:

> The quality of surface water varies widely because of both natural effects and human activity. In the upper basin, where the Bear River enters Utah from Wyoming, water quality is considered good. Water temperatures are low, as are TDS (total dissolved solids), alkalinity, electrical conductivity, hardness and sulfates. But the quality deteriorates as the river flows downstream. Return flow from irrigated land, sediment, animal wastes, municipal and industrial wastewater, natural saline springs, agricultural chemicals, and warmer temperatures combine to cause water quality problems in the lower basin. In general, each tributary stream shows a similar pattern of downstream deterioration, although some are much better than others.

Pollution is characterized as either point source, which comes from a specific site which typically requires a discharge permit, like a wastewater-treatment plant, or nonpoint, which is harder to locate, monitor, and control. The sources of nonpoint pollution are a product of human activities such as agriculture, construction, mining, recreation, urban runoff, channel modifications, and forest

Cache Valley residents sometimes use the riverbanks as a dumping ground, preferring not to pay the tipping fee at the county landfill.

Industrial operations, like the Jack B. Parsons aggregate unit at Dingle Bottoms in Idaho, create potential problems for runoff into the river.

management, and they are the primary generators of pollution in the Bear River watershed. Moreover, nonpoint pollution is a problem on both the main stem of the Bear and several tributaries: the Cub River, the Little Bear, and Spring Creek in Cache County; Deep Creek and Battle Creek above Preston, Idaho; and the Malad River in Box Elder County.

Without a doubt, grazing, dairies, and meat-packing operations contribute the most pollution in Cache County. The E.A. Miller packing plant and former operations like Tri-Miller were once heavily penalized. Materials drained from their slaughterhouses into ditches emptying directly into the Bear River.

"The guy I talked to in the Health Department said the ditches were red [with blood]," Mesner says. "He would contact the state, and the state would come up on their schedule, sometime Monday through Friday, and wouldn't capture it. Eventually he went out and took photos."

Many of these problems have been ameliorated, but they still crop up. In May 2005, the Utah Water Quality Board cited Ritewood Eggs in Lewiston for spilling an estimated two million gallons of chicken waste into an irrigation ditch feeding the Cub and then the Bear River. The company breached the dike of a holding pond at its composting facility while trying to move the polluted water to a second pond due to heavy runoff from rain. It didn't have a permit or an emergency plan to cope with the spill.

Now the state is focusing on nonpoint pollution, especially manure running into the river. According to Alice Lindahl, most early damage was due to animal feedlots between Trenton and Amalga in Cache County. The feedlots were designed in the 1930s to be on the river so that ranchers could dump feces and cow bodies cheaply. Live cattle create a problem, too, when ranchers allow them to graze right up to the riverbank, where they destroy riparian vegetation, thus increasing sedimentation of the river, adding to turbidity, and augmenting phosphorus loads.

And it's not just cattle polluting the river. Lindahl says that some Cache County residents have gotten into the bad habit of using the Bear as a dumping ground, rather than paying the tipping fee at the county landfill. It's hard for farmers, too, to wean themselves from using the Bear to deal with their offal.

"The people who work on the Bear tell me there's all sorts of illegal pipes coming out of milking parlors that are under overhangs and vegetation," Lindahl says. "There's still plenty of illegal dumping going on out there."

Pipes draining phosphorus and nitrogen, as well as pesticides, from fields create problems for downstream users. Pollution also gets into the Bear when ditches carry in the overflow from flood irrigation.

Increasingly, industrial operations like gravel mining are finding their way to the riverbanks. Surprisingly suburban encroachment poses an increasing threat, too, as more and more wannabe country squires purchase ranchettes next the river.

Five-, ten-, and twenty-acre ranchettes like these in Bear River City, Wyoming, do more than subdivide wide open spaces with barbed wire and vinyl fences. Their septic systems leak into the river.

"Right now I'm interested in these ranchettes, these five- to twenty-acre [plots] that people call 'weedettes'", Nancy Mesner says, "where someone has this dream of going in and buying this piece of rural paradise. So you have people dividing up agricultural land into these small ranchettes that aren't managed very well. It's causing a lot of odors. We have septic-tank problems all of a sudden, and everyone wants three horses and two cows. We've got noxious, introduced weed issues. So we are going to see our focus changed in some of these developing areas from agricultural impact to suburban."

That includes the increasing use of irrigation canals for storm-water collection. Urban and suburban areas are degrading both irrigation water quality and farmers' abilities to deliver and maintain their systems. "These systems are doing both irrigation and storm-water collection," Mesner explains. "If you think about it, an irrigation system is a delivery system that gets smaller and smaller as you get farther out in the distribution system, as opposed to a storm-water system that gets bigger and bigger. They are at cross-purposes, plus you get increased amounts of pollutants from urban areas draining into these irrigation areas."

Farmers point to these issues and say that it's unfair to single them out, that pollution to the Bear River watershed comes from multiple sources. "That's been a little bit of a sensitive issue among

(bottom left) Drought and periodic draws of water for downstream irrigation have made lakeside property at Bear Lake less valuable.

the agriculture community because it's seen as a put-down, but when you look at the land uses in the watershed, it's primarily agriculture. So it's a good place to start," Mesner concludes.

Hydropower Relicensing

Historically nutrient poor and oxygen rich, Bear Lake's famous turquoise transparency is threatened as water from the Bear River is stored in the lake, bringing with it pollution from upstream agricultural operations.

In 1999 PacifiCorp filed a hydropower license application with the Federal Energy Regulatory Commission to renew the power company's right to generate electricity at the Soda, Grace/Cove, and Oneida operations. When PacifiCorp undertook its review process in 1996, it reiterated the historical position that it was a caretaker of irrigation rights and had to honor those contractual agreements under the Bear River Compact. Moreover, that subordinate position of the power company was affirmed when Scottish Power purchased PacifiCorp and had to pledge to maintain the irrigation rights. While some in the watershed hoped to break the lock of irrigators on the river during the relicensing negotiations, it was generally understood that provisions of the compact would have to be central to the new license.

But the power company also tried to argue that the Bear River was a degraded environment and PacifiCorp had had minimal effect on its already-poor condition. The company maintained that since there was no way to improve the river, it should be allowed to continue its operations as before with no mitigation of environmental problems. Other players in the watershed quickly challenged that position. Things had changed over the past fifty years, and new interests demanded recognition of their claims to the river.

As land becomes more valuable in the New West and the norm moves toward itinerancy, modular housing crams the riverbanks in Evanston. The Bear is channeled supposedly to mitigate damage from flooding, but that only creates more problems downstream.

Besides the traditional federal agencies that would be asked to comment on the relicensing application—the Forest Service, the FWS, the National Parks, the BLM—other groups intervened. The Shoshone-Bannock tribes wanted to make sure their hunting and fishing rights under the Fort Bridger Treaty of 1868 would be honored. Idaho agencies stepped in as well: the Department of Environmental Quality, the Department of Fish and Game, the Department of Parks and Recreation, and the Department of Water Resources. But it was actually the nongovernmental organizations that forced PacifiCorp to make some begrudging compromises. The Idaho Council of Trout Unlimited, Idaho Rivers United, the Greater Yellowstone Coalition, and American Whitewater demanded that recreation and environmental mitigation be part of the new agreement, even though that meant that PacifiCorp would have to pledge money to repair the environment and cut back the approximate 84.5 megawatts of electricity it had generated in the past.

The FWS reported there were no endangered species in the stretch of the Bear River watershed affected by the PacifiCorp license, but it did note populations of Bonneville cutthroat trout, a species of special concern, in higher reaches of some tributaries. Fishers asked that any relicensing agreement require PacifiCorp to recognize that fact.

In the 2002 settlement agreement, PacifiCorp acknowledged that irrigation, cattle grazing, and hydropower development had led to ecosystem degradation and agreed to restore native fish habitat through mitigation funding. Furthermore, the company said it would conduct studies leading to the development of a Bonneville cutthroat restoration plan. That could include allowing the spring-fed water in Black Canyon above the Grace power plant to bypass the Cove forebay and restore water to a largley dewatered section of the river, which could become a prime fishery. In 2005 the pledge to conduct studies led to creation of a minimum flow in the bypass section.

Boaters had asked to have more water be released both below Oneida and, especially, through Black Canyon. Weekend releases would enable them to enjoy whitewater kayaking, canoeing, and rafting on their days off. That didn't sit well with fishers since fluctuating water flows are hard on fish and invertebrates, which can become stranded. Fishers feared increased flows could harm the spring-fed fishery in Black Canyon, besides confounding Bonneville cutthroat restoration. Aquatic ecologist Jeff Kershner voiced the frustrations of the fishers:

"I heard one of the fellows from the Salt Lake City groups, who said, 'Well, we want to turn on the water at ten, and we want *X* number of cubic feet per second to go down the canyon until four, and then we want to turn off the canyon at four because that's when we like to get off the river and have our cocktails.' I think my comment at that meeting was, 'If you want Disneyland, go to Disneyland. They have things like Big Thunder Mountain, and you can get in a tube and go down the thing, and go 'whoopee!' and you're not going to affect anything. But we're dealing with streams here. Streams are still streams. We're dealing with aquatic communities here.'"

"I find it curious," Kershner continues, "that some people that call themselves environmentalists are environmentalists until they want to do this recreation thing: 'Hey, we can go up on the Bear and it's only an hour and a half from Salt Lake City and turn that into our playpen.'"

Eventually PacifiCorp listened to the warring parties and offered a compromise. With irrigation rights still supreme, the company would redivert water from its pipeline when it became available and send it down Black Canyon for boating during specified periods of spring and early summer. It would notify the public when the releases would occur. But the company would try to honor the natural hydrograph of the river so that species would find high water when they were evolutionarily used to it; it would also monitor the effects of high water on native fish habitat in Black Canyon. No extra water would be released below Oneida Dam for boating since relicensing studies showed that current flows were sufficient to maintain the Class-One or -Two whitewater boating opportunities for beginners for about half of the summer weekends.

Last, PacifiCorp agreed to prepare a Historic Properties Management Plan to protect cultural resources and said it would require its land lessees to maintain a buffer zone on parcels near reservoirs to reduce grazing impact on riparian vegetation. PacifiCorp agreed to fence off cattle from sensitive land it owned around Cove. After the creation of an Environmental Coordination Committee composed of all the intervenors, everyone signed off on the 2002 agreement. All parties got a little something besides recognition, and PacifiCorp got its license for another thirty years, although it had to pay more than it had anticipated. Times had changed.

Wild and Scenic Rivers Act Controversy

In 1968 Congress passed the Wild and Scenic Rivers Act. The first section of the Act states, "It is hereby declared to be the policy of the United States that certain selected rivers of the Nation, which, with their immediate environments, possess outstandingly remarkable scenic, recreational, geologic, fish and wildlife, historic, cultural, and other similar values, shall be preserved in free-flowing condition, and that they and their immediate environments shall be protected for the benefit and enjoyment of present and future generations." The act also created a National Wild and Scenic River System. Rivers could be included in the system by Congress or state legislation if a governor reached agreement with the secretary of the interior.

Rivers under consideration have to go through a two-step process. The first is eligibility and classification. The second, and more politically difficult, is suitability. Eligibility requires that a river be "free-flowing" and have "outstandingly remarkable" value. These values can be scenic, recreational, historical/cultural, geologic, fishery, wildlife, or hydrologic. Rivers meeting eligibility criteria are then classified into one of three categories: *wild,* describing rivers with reaches free of impoundments and accessible only by trail, whose watershed or shorelines are essentially primitive and whose waters are not polluted; *scenic,* describing rivers free of impoundments and accessible in places by roads, whose watersheds or shorelines are largely primitive and undeveloped; or *recreational,* describing rivers that may have undergone some development, impoundment, or diversion in the past and are readily accessible by road or railroad.

Determining suitability for wild and scenic designation relates to practicality. Suitability analyzes the area's values in the context of current and possible future ownership and uses. This segment of the process looks at the environmental, economic, and social impact that wild and scenic designation will have on these uses. It queries public-agency interest and can suggest adjacent land acquisitions. Suitability must compare management alternatives and solicit public review through the National Environmental Policy Act.

Cascades on Stillwater Fork, one of the Bear River tributaries for which Wild and Scenic status is sought.

In 1995 a team from the Pocatello office of the BLM surveyed the Bear and its tributaries in Idaho. It identified several reaches of the Bear as having "outstandingly remarkable values." All were classified as "recreational" for their geologic, hydrologic, and wildlife values. Several tributaries had Bonneville cutthroat trout that could be protected. The classification placed a moratorium on new mining claims and hydroelectric power projects, but it could only last for six to eight years, according to the act.

The Forest Service once decided that the Stillwater Fork of the Bear had wild and scenic values. Later, the URC petitioned Representative Jim Hansen to propose twenty-three miles of the Bear River in the Uinta Mountains for Wild and Scenic status: Left Hand, Right Hand, and East Forks. But Hansen turned down any wild and scenic designation in his district because it precluded resource development. To date, Utah has had no elected representative championing the wild and scenic cause.

The Politics of Planning

A major problem with responsible water development on the Bear is the number of political constituencies involved. If all politics is local, water planning is the best example.

Nancy Mesner cites Cache County. The county has always had a lot of water due to abundant groundwater, so water planning has never been a big issue. In fact, plans for a water conservancy district in Cache Valley were shot down.

"There's a lot of water and this [county] is kind of an outlying area where folks don't want to be dictated to," Mesner explains. "Planning is such a dirty word up here, more so than along the Wasatch Front. These are tiny little towns that like to function like little kingdoms. They don't want to work with each other in a very seamless way."

Cache County voters also were concerned about the lack of democracy in the water-conservancy district. Since district board members are not elected but appointed, governance became a factor. "People saw it as another taxing authority without any representation," Mesner says. Both ends of the political spectrum united, proving that you can favor planning yet still oppose being dictated to by officials you didn't choose.

David Ovard gets frustrated with state water politics. He'd like Utah to pass a bill creating a new model for land development that requires certain standards of water conservation. But real property-development interests are powerful. "They [state legislators] all get really cold feet," Ovard observes, "when it comes to legislation."

And while Ovard finds environmental organizations helpful in their support of conservation, they don't feel the JVWCD is doing enough. From his perspective, Ovard maintains that environmental groups are politically fractured. He divides them into consumptive and nonconsumptive.

"You have people that like rivers. You have people that like birds. It's just a wide range of environmental interests," Ovard explains. "We were criticized on the Bear River project. They said, 'Before you go to the Bear River, you ought to do everything you can to develop local water supplies.' This was the [Utah] Rivers Council. So we said, 'All right.' And we started scratching our heads, and we said, 'There is an untapped resource here: shallow groundwater.' So we proposed a series of shallow wells and treatment plants to utilize this unused untapped resource. Then you have the Audubon Society and the Friends of the Great Salt Lake saying, 'Hey, you can't do that. What's the impact going to be on wetlands? What's it going to do to the Great Salt Lake?'"

"So, when you start to do a water project," Ovard continues, "you have to be prepared that you're going to be hit from every side, from every group's special point of interest. Environmental groups can be helpful in helping to shape public attitude and point out the need for environmental values. But, when they're pursuing their own agendas, they can be very self-interested. For any water project to be successful, you have to bring in enough scoping process, public involvement process to allow everybody to give their input. Then, hopefully, public policy and public need will dictate what needs to be done."

Zach Frankel, former executive director of the URC believes that the costs of diverting the Bear River for water development is always going to be a problem for state legislators, and the recent decision to focus first on a pipeline bringing Lake Powell water to more rapidly growing southern Utah tends to support that position. The other political force he thinks will kill Bear River water development is the National Environmental Policy Act. Any diversion of water requires an environmental impact statement and approval from the EPA. The law also requires the diversion to be the least expensive alternative to the environment. To measure that claim, the EPA can reference twenty years of successful water-conservation programs in Denver. That city has been able to provide new water by retiring and purchasing irrigation water rights. Utah hasn't even begun to investigate that possibility. Consequently, until Utah shows that "least expensive" option has been exhausted, the EPA probably won't approve diversion of the Bear.

Alice Lindahl is less optimistic, feeling that the urban water-conservancy districts are still locked into a dam-building mentality. She believes the politics of Bear River water development are being driven by the urbanization of the West.

"There are people around here...water is kind of like a religion, and building dams is like a mantra," she says. "There's no such thing as a bad dam. These are people who remember the thirties when the feds came out here and built dams for free. Land that was not valuable became, overnight, very valuable. Most of them are real estate developers who are pushing the dams now because their land can then be turned into cul-de-sacs. They claim all of the farmers really need this, but farming is on the way out. It's really residential—suburb development."

Referring to the promise in the Bear River Development Act for water for Cache and Box Elder farmers, Lindahl continues, "The state gives a subsidy for dam building for the people who are going to use it for agriculture, but they'll just turn around and sell it to the developers. No one even pretends that they're really doing this for agriculture, even though on paper it looks like it. They [developers] will give it to them [the farmers] until the cities need it, and then they'll [cities] make an offer they [the farmers] can't refuse."

Cobble-and-gravel-bottomed riffles below Oneida Reservoir oxygenate pools where trout feed.

Ruler straight and falling fast
Channeling scrapes my bed away.

Give me back my bends,
And old mysteries revive
In duet of riffle and pool.

Wrap my silty banks
With rock,

Plant shoots of willow and
cottonwood,

Let sun and water energize.

Where shade protects
My fluid edge

I give you life
In return.

PacifiCorp's restoration of wetlands around Cutler has provided new habitat for pheasants and better opportunities for pheasant hunters to access the islands in the reservoir.

Chapter 7

Mitigation on the Bear: Repairing a Century of Misuse

Sometimes it's hard to believe there has been any mitigation of damage caused by human activity on the Bear River. Nevertheless, concerned citizens from the environmental community, government, and the private sector are working on improving water quality and habitat. It's just that, measured against a century of thoughtless degradation, this tender awakening of public responsibility is in its infancy. Nancy Mesner, Assistant Professor in the Department of Aquatic, Watershed and Earth Resources and Utah State University's water quality extension specialist, points out that there are only twenty years of data measurement.

"Cases are being made now that we are seeing some water-quality improvement on some of the smaller subwatersheds," Mesner says. "There are some other cases that could be made that we are seeing some degradation in the main stem of the river. What I see is kind of a cloud of data" with no clear trend. Nevertheless, Mesner feels that the 1996 Utah State Water Quality Management Plan, a guidance document for identifying and addressing problems of water pollution, was a step in the right direction. "I think we are *paying* more attention toward water quality," Mesner continues. "I think we are putting more money into managing our rivers, and in the last few years, there's been a lot more emphasis on doing a better job of monitoring the response to the river, where they've put a lot of money into restoration activities."

Because livestock operations have such immediate and identifiable impact, many remediation activities are focused there. "We're trying to build manure bunkers on the tributary streams going through little dairies," Mesner explains. For instance, Bear River Resource Conservation and Development, a nongovernmental agency, has been working with Cache County ranchers to stop manure from working its way into the river. The Gordon Zilles dairy on Hyrum Slough produces three thousand tons of manure annually. Once it ran off into the slough. But recognizing the problem and committed to mitigation, Zilles built three manure bunkers on his property. He cleared the slough of debris and trash and now pumps liquid manure runoff to a storage bunker to be used later to fertilize fields.

"We're also trying to increase people's awareness of the application of manure," Mesner continues, "so that it is being applied in

At the Gordon Zilles dairy in Cache Valley, concrete bunkers prevent manure from flowing into Hyrum Slough and then into the Little Bear River. Zilles pumps liquid manure runoff to a storage bunker to use later for fertilizer on his fields.

ways that the plant can actually use it, rather than just when they [farmers] wanted to."

"How is manure spreading a problem?" I ask.

"They spread it on pastures and croplands," Mesner replies, "and that's a good use of it because it's recycling it directly into your system. But if you put it on frozen land or on ice, or if you don't till it in, or depending upon the season in which you're applying it, it is more or less incorporated by the plant."

"The other thing that was a problem in applying it," Mesner continues, "was that people weren't being careful about the spatial patterns. They'd haul it twenty feet and dump it"—and Mesner laughs—"because it's hard to manage. People were viewing manure as a waste rather than a resource. And that's where a lot of the effort has been: trying to get people to see manure as offsetting the cost of fertilizer. And improving the structure of the soils that fertilizers won't do."

Nonpoint source pollution of the Bear is more difficult to mitigate "because you're limited by participation," Mesner says. "All the nonpoint source stuff is voluntary. It's not mandated."

The politics of nonpoint-source pollution are more intricate and convoluted than point source pollution, even if the latter has a more powerful economic constituency like dairy farms and slaughterhouses. At least with point source pollution, you can determine

One way to minimize problems from livestock drinking from the river is installing fencing in a configuration that allows the cattle to reach the water without doing undue damage to the banks, the way it's done on this ranch in Dingle Bottoms in Idaho.

the origin. Mesner offers an example: "Once you have a nitrogen molecule in the water, whose was it? It's difficult to truly identify the sources. So we tend to focus on best-management practices because, if everybody is following best-management practices, then we see a benefit to the stream, and no one has to be identified individually."

Mesner itemizes some of these best-management practices that are slowly being adopted by farmers in the Bear River Basin, especially on the bluffs overlooking the river in Cache County. Rather than furrowing land immediately after harvesting, farmers can do anything from not tilling, where you just reseed, to being more judicious with the way you till. The goal is to keep the soil on the land. Farmers try to maintain a vegetative cover so that if there is sheet-flow runoff, it doesn't take the soil with it. The plants force the water to soak in. This is conservation tillage.

Contour farming obeys the same principle. When farmers till fields, they try to follow the contours of the land instead of clearing a landscape from a high elevation straight down, which simply creates a pathway for water and sediment to move directly off the field. Tilling with the contour creates a series of tiny berms that capture the runoff, slow it down, and direct it into the ground.

In strip cropping, a farmer alternates several types of crops. One that needs high phosphate can be planted one year, and a low phosphate crop the next. Also a farmer can plant different kinds of crops or provide a buffer area between crops to stimulate bird activity, drawing in birds to take care of pests. Any of these procedures minimizes runoff of pesticides or excessive phosphorus into streams.

Sometimes farmers plant crops during the winter, instead of tilling the field, to stabilize the soil. These are cover crops. Farmers practice grade stabilization to avoid irrigating a field that is too steep. Too steep a grade creates pull pressure that sucks moisture out of the soil and destabilizes slopes. Planting vegetative strips immediately below a field works like a wetlands. The riparian vegetation captures any pollutants in the runoff and helps filter them through the soil.

Only a minority of enlightened farmers, though, have embraced best-management techniques. Sometimes their resistance has to do with inertia and tradition. At other times, they consider that these practices complicate the economics of farming. But in the long run, farmers who protect their soil are going to be farmers in the future. They just need to look further into the picture.

"I'm always flabbergasted," Mesner says, "when I go out to places and see fences hanging out over the river. To me, I look at that and see the failure of somebody to hold onto their land. Your field is gone. But you see a lot of resistance to measures that would hold onto those banks, like keeping a riparian buffer strip, because they feel like they are losing some of their cropland."

PacifiCorp and Wetlands Mitigation

PacifiCorp has always been a major player on the Bear River through managing irrigation water rights at Bear Lake and generating power. Increasingly the company is becoming a force in mitigating damaged habitat due to farming, power generation, and simple geomorphic processes. "The power company has made a lot of effort to improve wildlife habitat on the parts they own," according to Alice Lindahl, biology lecturer at Utah State University. "So those parts really look pretty good, although they weren't as badly damaged [as other areas]."

When PacifiCorp faced relicensing for Cutler Dam in 1995, the company knew it would have to take measures to improve the habitat around its operations. Suspended sediments had always been a problem, and Cutler Reservoir was filling in. There were increased phosphorous and nitrogen levels, largely due to farmers tilling to the edge of the reservoir and making no attempt to control runoff. Bacteria counts were up due to feedlot operations. Vertical shoreline banks were sloughing into the main stem of the Bear, its tributaries, and parts of the reservoir. Most vegetative cover was gone. Wave action and water-level fluctuations were eroding the shoreline, and concrete and car bodies used earlier to halt that erosion were creating aesthetic problems. Grazing had limited the height of cover. Poor water quality had inhibited submerged plant and invertebrate production, making food resources scarce.

One advantage PacifiCorp had was that it leased much of the land surrounding the reservoir to farmers. So it terminated those leases at the end of 1993 and created new ones with covenants based on sustainable and beneficial resource-management goals.

PacifiCorp had an added advantage in knowing that power generation at Cutler was more uneven than at other plants on the Bear. While its turbines can generate 30 megawatts, there is effectively no water for generation during the irrigation season from April to November. Because Cutler was less important as an overall system producer, PacifiCorp could implement mitigation plans that would be more expensive in lost power generation at other hydropower sites.

Moreover, with sedimentation due to silt from Lake Bonneville filling in the reservoir, building enough head for power generation was an ongoing problem. Most water has to be stored at Bear Lake for that reason anyway. A one-foot drop in elevation at Cutler lowers the surface acreage from 5,500 to 3,000 and the acre-feet capacity from 13,200 to 5,200.

"It's so shallow that you have to know where you're going in a canoe in order to avoid poking a mudflat," Lindahl says. So PacifiCorp could afford to be magnanimous in its proactive approach to repairing damage and project good corporate citizenship. The company was less willing to take similarly aggressive mitigation measures during the relicensing process for the more efficient Last Chance/Grace/Cove/Oneida operations upstream. "Nevertheless, it's been a real help," Lindahl admits.

PacifiCorp tackled the job at Cutler with multiple measures. It decided to moderate water fluctuation to improve surrounding cover by not inundating it and also to reduce erosion. The company required irrigators to notify it at least one week in advance of their diversions so that it could better manage that fluctuation.

PacifiCorp planted vegetative buffer strips of medium shrubs and trees between reservoir wetlands and adjacent fields to build a canopy and encourage biodiversity. Moreover, these vegetative strips helped prevent nutrients from entering the reservoir from filter-sheet runoff. To protect the buffer strips from wandering livestock, the power company erected new fencing.

By signing the new leases, area farmers had access to most of their original acreage, minus the strips used as buffers. But the new leases prohibited tree cutting, grazing, burning, and herbicide spraying in critical habitat areas. Previously grazed leases were treated for noxious weeds and then reseeded with native grasses and forbs, interplanted with shrubs and trees like native willows, cottonwoods, and dogwood.

When PacifiCorp granted new cattle-grazing leases, it required that they be of short duration with high-intensity management and a rotational pasture system. PacifiCorp splits the proceeds from the hay and grain harvest with the lessees, and if Canada geese, sandhill cranes, or other waterfowl lessen that output, those deficiencies are deducted from PacifiCorp's percentage of the yield. In some areas, PacifiCorp has planted fields with decoy cereal grain and created pastures to reduce crop depredation on adjacent farmlands by Canada geese.

Although fishing on Cutler Reservoir never will be spotlighted in outdoor magazines, PacifiCorp has tried to increase populations of black crappie and channel catfish in deeper areas by creating fish habitat structures. The power company replaced the car bodies and concrete with sloping and riprap, improved public access, and developed new recreational facilities. Hunting on lands or islands in the protected area is still allowed, but not camping. Finally, PacifiCorp pledged to monitor the health of all the newly planted vegetation in the buffer strips and reseeded areas, besides expecting to see increased numbers of wildlife due to improved habitat.

PacifiCorp also owns lands adjacent to the Bear River upstream from Cutler Reservoir called the Bottoms. It purchased this land from farmers by condemnation when it needed to create a flood zone to manage water in Cutler Reservoir. Of course, that created a lot of animosity between the power company and farmers.

When that land was in private ownership, it, too, became environmentally degraded. It was cleared, bermed, leveled, and farmed. Farmers tried to drain the wetlands to add acreage for cultivation. Woody vegetation good for upland birds was taken out, and grazing prevented reseeding. What once was a rich environment for bunch grasses or forbs became a sod-forming grass monoculture. The multilayered canopy was chopped down, and with it, songbird habitat.

PacifiCorp leased the land to the Utah Division of Wildlife Resources (DWR) in 1988 for fifteen years. Their mutual goal was to increase wildlife population and recreational use. They used disturbance tillage to break up the sod grasses and created corridors for wildlife movement. The partners agreed to plant corn sorghum for winter feeding and arranged with adjacent farmers to leave one-acre plots of standing corn. Together, these new measures provided winter forage for songbirds, which had been lacking, and lured deer away from pastures and stacked hay.

While this has evolved into a positive mitigation program, PacifiCorp has been somewhat disappointed in the DWR's management. It is considering groups like the Bridgerland Audubon Society, Ducks Unlimited, Pheasants Forever, and the Bear River Watershed Council (BRWC), a coalition of nongovernmental organizations in northern Utah and southern Idaho, to take over some of the management. It's not that the state of Utah lacks knowledgeable personnel. The problem is due more to the politics of conservation and taxpayer economics. Jack Greene, the recently retired environmental science teacher at Logan High School, offers his analysis:

"I'm guessing that it's really rough," Greene says. "DWR—they get crunched, they get hit. They never know from year to year what their budget will be, what the politics are going to allow them to do. I know a lot of the DWR field people, and they're great people. They have good intent. Very capable individuals. But when you constantly get new administration, and not know from year to year what your budget priorities are going to be, it makes it really tough to do."

Greene and other environmentalists like Alice Lindahl and Val Grant, president of the Bridgerland Audubon Society, are encouraged by PacifiCorp's proactive stance on river mitigation, largely because of Eve Davies, a trained biologist who was hired by PacifiCorp to direct the company's environmental efforts on lands it owns in the Bear River Basin. Davies, who used to teach at Utah State University, was charged with developing mitigation plans for Cutler. As part of that overall plan, Davies worked with the Bridgerland Audubon Society to arrange for its purchase in 2004 of 150 acres that PacifiCorp owned in the Barrens area. A kind of capstone to the decade-long restoration of the lands around Cutler, sale of the acreage ensured some protection for migrant sandhill cranes, besides complicating the choice of Amalga/Barrens as a potential dam site on the Bear River.

River Revetments in Wyoming and Utah

Any river, the Bear no less, hates to be channeled; to become a straight line is as unnatural for a river as imprisonment is for an animal. A river must wander, its meandering a natural response to its need to dissipate energy. Evanston, Wyoming, found that out when it analyzed the severe problems that channeling the Bear River had created.

Some see channeling as a flood-control measure. In reality, channeling moves the flood problem downstream to someone else. It's a losing cause because a river will try to regain its meanders by eroding its banks. Before that happens, the increased velocity caused by channeling causes the downward erosion of the streambed, producing a steeper grade that wipes out riffles and pools that provide habitat for fish and insects. A channeled river becomes a dead river.

When it channeled the Bear, Evanston halved the length of the river as it flowed through town. When it tried to remedy the damage, the city was limited by private ownership of lands next to the river. Nevertheless, it hired engineers to put some meander back into the river and create river structures or bank revetments to mitigate other environmental damage.

Evanston faced other problems, too, when it tried to rehabilitate the Bear. Besides losing pools, the river either braided or had become overly wide with shallow channels. Studies showed increased nonpoint-source pollution due to urban encroachment

Jetties can direct water to deeper pools during low flow to maintain aquatic habitat and a healthy fishery, while still allowing the river to scour silt from the pools.

by homes and industries. To maximize their acreage, these businesses had cut down trees and shrubs along the river, and the resulting lack of shade and protective cover effectively eliminated terrestrial food input into the river system.

The consulting engineers began their mitigation by figuring proper pool-riffle ratios and sequences for natural reaches. They built desirable meander wavelengths into their plans, all with the object of reversing the accelerated flow. They added pools at the concave bends of the meanders and point bars on adjacent convex bends. To make sure that pools wouldn't dry up during low flow on the Bear, the engineers used boulders to deflect water into them. They added V dams, drop structures with the point of facing upstream, to create drop pools to dissipate stream energy. The engineers put bars in logical places and constructed jetties to concentrate low flow, thus maintaining water depth in pools while allowing the river to scour silt from the pools during high flows, which also discouraged beaver activity. Riffles were roughened with medium-sized boulders to stabilize them and provide habitat for macroinvertebrates like caddisfly and stonefly larvae. Downstream from the riffles along the banks, where the river

Riprapping helps to stabilize streambanks against the erosive action of the river.

formed back eddies, engineers built fish nurseries. Later, the fish would establish feeding lanes as they picked off larvae drifting downstream from the riffles.

Repairing the river channel wasn't enough. The engineers also had to revive the riverbanks. They stabilized them with two rows of boulders and used brush layering to anchor the soil. Along the riverbanks and on newly formed point bars in the river, they used stake and bare-root plantings of cottonwood and willows to restore the canopy and bring the river and its natural stakeholders back to life.

Utah's two demonstration projects are on the Cub River and the main stem of the Bear. Environmental engineers have ramped up the banks from their former sheer vertical slopes and planted vegetation to stabilize them. In other areas, they've planted live willow branches and other vegetation in the riprap material on the banks so that it takes root, as well as putting some structures in the river to dissipate flow energy.

There's also an operation to fence off a dairy from the Cub. But the issue becomes how to provide water for the animals, which can be costly for farmers because lifting water is expensive. The usual solution is drilling a well or putting in a diversion structure.

Sometimes revetments like this riprapping in Dingle Bottoms are ineffective for riverbank protection during periods of drought when fast, low-water currents undercut them.

Box Elder County's Wetlands Master Plan

Box Elder County, Utah, is taking an aggressive stance to balance development with wetlands protection. Knowing that the Clean Water Act and the Rivers and Harbors Act give the Army Corps of Engineers regulatory control over wetlands impact, Box Elder doesn't want to find itself in protracted battles over wetlands mitigation resulting from urban development. The corps has the power to require a permit to discharge dredged material into U.S. waters and may demand compensatory mitigation of impact, including restoration, enhancement, and/or preservation of existing wetlands or creation of new ones. So in 1997, the county created its Great Salt Lake Ecosystem Plan steering committee. Its mission was "to conserve and enhance the integrity of the Great Salt Lake wetland ecosystem in Box Elder County, incorporating provisions for appropriate urban development, infrastructure needs, resident livelihoods, and quality of life, while ensuring perpetuation of these important natural resources."

With consulting engineers, the county drew up a list of conservation and urban-development goals. By showing a good-faith plan to balance those goals, the county could streamline any future permitting process from the corps for wetlands impact. That would require developing a Special Area Management Plan (SAMP). If the plan was approved, the corps would grant the county a general permit that would provide prior approval for some development projects if they recognized legal requirements for wetlands protection or mitigation. This easing of the regulatory process would lessen the need for mitigation efforts and provide landowners with an incentive to conserve wetlands on their property.

After an inventory of the county's wetland ecosystem and socioeconomic needs, the steering committee listed conservation planning goals that would lead to a "desired future condition" for Box Elder County's wetlands:

- Establish an interconnected system of wetlands, rivers, riparian areas, and uplands that preserve wetland functions and values.
- Emphasize protection, enhancement, and restoration of existing wetlands over the creation of new ones.
- Ensure no net loss of wetlands and their values.
- Protect the existing hundred-year floodplains of Great Salt Lake and Box Elder County rivers and their ability to handle flood flows in ways that prevent or minimize hazard to public health safety and welfare.
- Improve the quality of rivers, creeks, and streams.
- Work cooperatively with landowners who have wetlands or tributaries to adopt best-management practices.
- When appropriate, incorporate public-education components and recruit public involvement.
- Provide for long-term maintenance, management, and monitoring of wetland projects to make sure they meet these goals.

The next step in the process was to list urban-development goals. The first priority was to simplify the Clean Water Act permit process by developing the SAMP that allowed impact to certain wetlands if it ensured no net loss of wetland function. The rest of the goals identified development priorities:

- Encourage urban development that achieves the "desired future condition" and minimizes adverse impact by creating acceptable zones.
- Identify, classify, and map urban-development and conservation property that is suitable for mitigating wetland impact.
- Ensure that the wetlands plan gives protection for land for growth equal importance with wetland conservation and preservation in benefiting current and future generations. The eventual plan should provide a mechanism for resolving conflicts that places equal value on wetland conservation and urban development.

Obviously, the last goal will become the proverbial sticky wicket. In our polarized society, it requires that competing sides recognize the values of the other side as equally legitimate. But as conservative or progressive ideologies become entwined in water politics, as in all other areas of contemporary politics, resolution too often seems tragically out of reach. This sad political state is tellingly reflected in a 2005 bumper sticker: "A Tolerant Person Is One without Convictions."

Water management in the Bear River Basin has been marked by fits and starts. Concerned conservationists created the BRWC to focus attention on the Bear River and the Bear River Range. Sometimes that bifurcated mission has led to conflicts in the organization. Nevertheless, the BRWC has been an active player in focusing public attention on the needs of the river. Moreover, its efforts to highlight ORV damage in the adjacent national forest can potentially improve water quality by reducing erosion, which winds up as silt in the river.

Cache County has been ambivalent about water planning. The county voted down a water conservancy district some years ago for two reasons. First, most county water comes from plentiful aquifers, so planning didn't seem imperative. Second, voters questioned whether forming a district would be an effective conservation measure.

"People saw it as another taxing authority without any representation, too," Nancy Mesner said. Yet in July 2005, the six-year drought and increased urban growth brought the issue to a head again, in spite of the bountiful water year. The Cache County Water Advisory Board listened to Larry Anderson, director of the Utah Division of Water Resources, tell them that, if they wanted to become players in developing Bear River water, they needed to have legal status. The volunteer advisory board lacked authority at the state level.

I ask Nancy Mesner about her opinion of such organized efforts to address problems in the Bear River watershed. "There's a lot of stuff going on," Mesner replies, "but I think the challenge is continuing to make sure that it's done in an integrated way. We have a tendency to slap things here and there and not being careful in making sure that where we do put our limited resources will have the most impact."

"I think we will continue to be a multiply used, multiply managed system," Mesner continues. "I would like to see increased coordination of that management. I think we're seeing that. We have the Bear River Compact and the Bear River Water Quality Task Force, which is now an advisory body to the Bear River Compact. So there's a level of cooperation and coordination that we've never seen before."

Mesner would like to see a restoration of the natural hydrograph of the river. She points out the ramping problems and vertical banking resulting from flow fluctuations at Oneida Narrows Dam. She also highlights the problem of old cottonwoods on other managed areas that aren't being replenished because there isn't flooding. She would like to see riparian areas restored, although conservationists have wondered for years how large these areas were before the dams were built.

"That is the million dollar question. Clearly this area was a big wetland at one time," Mesner says. "There are drain fields all through Cache Valley. We've tilled it and drained it. I don't know what it looked like. I don't think we're ever going to restore it to pre-European conditions. Even if it were desirable, it's not possible. But I think we can certainly make it function better."

"I think the Bear River is such an untapped resource for recreation," Mesner continues. "People don't realize what a gorgeous river it is. When you're on the river, there are a few of these sort of…nasty spots that everyone's has seen pictures of—in the low gradient areas, we have some real problems—but it's such a beautiful river. And I think just that alone makes it a real resource."

Clarity of light accompanies an approaching storm in Dingle Bottoms.

Worn myths
Dim your eyes
To sins that profligacy wreaks
On you and me.

An illusion of desert water,
Refracted through consecrated truths,
Reveals

Life out of balance.

I offer the miracle
Of lake and snow

If you save
The fasting time
To map the course
That we must take.

The Bear River Development Act requires that, at some time, northern Utah must receive 120,000 acre-feet of water from the river. The most politically and environmentally feasible dam site for that water is Washakie, just west of Portage, Utah. The reservoir would fill the shallow depression on the horizon by pumping and lifting the Bear River to transfer it off-site, and out of its course.

8

Chapter

A Stark Future for the Bear River — and Us

The facts are stark, and the disconnect is obvious. Utah is the second-driest state in the nation, yet it is also the second highest in per-capita water use. That profligate use of water in a desert environment is stultifying. The message it sends is foreboding.

According to a March 2005 study by the Bureau of Economic and Business Research at the University of Utah, residents of Salt Lake County consume 288 gallons of water a day, 18 percent more than the average 243 gallons in the Intermountain West. Logan consumes a whopping 370 gallons per person. Citizens of Idaho and Wyoming, which have more water available, actually consume less. But no state in the Intermountain West has an unlimited supply of water, and few consume it wisely.

If you look at how water is used, the breakdown is more disconcerting. Approximately 30 percent goes to municipal and industrial use. But about 60 percent of that amount is used to irrigate Kentucky bluegrass during the arid summer months. In states with a relatively large amount of acreage devoted to agriculture, most developed water, more than 70 percent, goes to farming. At first glance, that appears good since we all have to eat, but in many farming areas in the Intermountain West, hay is the principal crop, livestock feed of comparatively low economic value which consumes a lot of water.

So, if you are a water planner and want to start thinking about conservation and better management, rather than falling into the old mind-set of developing new supplies, usually with a dam, you can look at ways to use agricultural or municipal water more efficiently. Perhaps more water should be redirected from agriculture to municipal use via market forces. Indeed, that already is happening.

But complications arise when it comes to rationing agricultural use. When water is taken out of rivers like the Bear and used to water crops, agriculture begins partially to take over a role that floodplains used to play. Water seeping into the earth from farm fields can recharge aquifers that cities use for drinking water, as New Mexicans found out in negotiations between Albuquerque and environmentalists over protecting the endangered Rio Grande silvery minnow and maintaining groundwater/surface flow equilibrium. A study showed that one-half of the aquifer recharge in the Albuquerque basin is due to seepage from Middle Rio Grande Conservancy District ditches and flood-irrigated fields.

Nevertheless, farming hasn't become a substitute for river flooding. While agriculture provides some recharging of aquifers, there is a net loss to overall surface water when water is diverted from a river and used for other purposes.

By sustaining agriculture, water helps maintain open space. If agricultural receipts and profits can't be sustained with relatively cheap water, then farmers and ranchers have no choice but to sell increasingly valuable land to developers. JVWCD general manager David Ovard would hate to see that happen, even though he indirectly competes with farmers for limited supplies of water.

"I love cultivated fields, and the lifestyle, and the smell," Ovard admits. "Agricultural life is a real asset. My personal feeling is that the state would want to hang onto as much agricultural as they can. From a pure economic sense, I don't know if I like the feeling of having to import all of our food from out of state. Whatever we can do on our own. Self-reliance. Independence."

Water Conservation

Although they are Johnny-come-latelies to water conservation compared to other western states, Ovard is prodding his agency staff and the state of Utah, trying to catch up.

"We kind of came on the scene late, but the last four years have been very aggressive," Ovard explains in a 2003 interview. The agency has contracted with Utah State University to provide water audits for homes in Salt Lake County. Ovard also touts the sixty thousand dollars in 2003 that was budgeted to advertise statewide water conservation, which doesn't sound like much when you consider it was divided among six water wholesalers like the JVWCD.

Ovard has great hope for another of his agency's initiatives: providing model water-conservation landscape ordinances, both commercial and residential. "In my mind, the control point for conservation is the city planning commission," Ovard says. "When a developer for a commercial development, or a residential development, or even individual homeowners—if the city would take it that far—if they come in and apply for a building permit, the city could tell them exactly what was acceptable in regards to landscaping to save water. That is the most cost-effective measure, period, that we've been able to identify. So Sandy has adopted our ordinance. West Valley City has adopted it for retail and easements served there. South Jordan has adopted it. South Salt Lake adopted just the park strip portion. Murray's adopted sort of a softer version."

Although its headquarters once was surrounded by acres of Kentucky bluegrass, JVWCD has torn up the sod and replaced it with six demonstration gardens. Even the thirstiest garden, which looks like the typical landscaped yard in Utah, would save homeowners 50 percent of their water use with minimal changes, showing just how much the average yard is overwatered. The water audits are helping, too, with Utah State University auditors reporting they save their residential clients an average fifty inches of water a year on their landscaping.

"Our goal is to change what is publicly acceptable for an outside landscape. We believe people can have a beautiful landscape and still save water," Ovard says.

"How are the developers reacting to those options at the city level?" I ask. "Are they saying, 'We're excited about that. We'd like to offer that'? Or are they saying, 'Our customers don't want it. They want the old way'?"

"It's largely aimed at the commercial initially," Ovard replies. "The residential is a little tougher. If you have a homeowners association or something like that, they'd fall under this. Some of the cities are a little reluctant to put that on individual homeowners. So we're still trying to gain acceptance for that."

"We're still going to have a lot of yards where we're going to have to figure out a way to help people retrofit and make them water efficient," Ovard continues. "But the state's population is going to increase by 70 percent over the next thirty years. That's a million and a half people. If we were able to get those ordinances in place and every new home and every new business had a water-efficient landscape, in thirty years our job is almost half done as far as reducing outside water use. If we do nothing, it's going to be twice as big a job as it is right now.".

"So that's my vision," Ovard explains. "It's time to create a new model. And every chance I get, I preach it. With legislative committees, with governors, I say, 'Let's do a bill. Let's tell all the cities, "This is what's desired." Let's have the state take some leadership.'"

"You're going right into my next question," I chuckle.

"But they all get chicken. They say, 'No. We don't think it's our job to tell the local governments what to do.'"

Utah state government has made some efforts to promote water conservation. Looking at ways to conserve water through irrigation practices, the Division of Water Resources suggested several alternatives in its 1992 Utah State Water Plan for the Bear River Basin. It recommended converting from gravity flow to sprinklers, which can reduce waste from overirrigation. This practice tends to diminish diversion of water from the head of a system, although it also can lead to overconsumption of water by some crops. Better management of irrigation water by scheduling the time and amount for optimum plant growth can also increase crop yields and farm income while reducing use. Also farmers are encouraged to line canals or switch to pipes to minimize water loss via seepage.

Utah supports residential water conservation through its "Slow the Flow" campaign initiated by former Governor Mike Leavitt and continued by his successors, Olene Walker and Jon Huntsman. There is some evidence that that is working, too. The Utah Division of Water Resources determined that water used in Salt Lake City dropped by 17 percent from 1995 to 2003 as residents responded to pleas to conserve water during the recent drought. But the rate of conservation has slowed in the past two years. Water officials are concerned about the staying power of conservation. Looking at the way Las Vegas residents backslid in their conservation efforts, conservationists worry that one wet water year like 2005 can change attitudes. While there is an initial big saving in water use the first few years of a conservation campaign, public interest tends to wane as people grow tired of the programs and the media stop covering them. In summer 2005, there were far fewer "Slow the Flow" public-service announcements broadcast on Utah television stations. Apparently the state and the media felt Utah's water problems were over.

Utah state water officials are calling for a 25 percent reduction in use by 2050. David Ovard and his agency are figuring a 25 percent reduction in use via conservation measures by 2025. Even with these figures, the JVWCD maintains that Bear River water will have to be developed. The URC believes even that goal is too timid. It agrees with Ovard on the 25 percent reduction by 2025, but it believes the state can achieve another 10 percent by 2050, for an overall 35 percent reduction by midcentury.

"Thirty-five percent is a feasible, but stronger, goal for the state, especially for the reductions we've seen since 1995," according to Merritt Frey, current executive director of the URC. "We see that Utahns have proven their willingness to step up to the plate, and we're trying to encourage the state to show more leadership."

The state responds by saying it wants an "achievable goal," one that is not set too high.

"But the definition of leadership is to step out in front and bring people along," Frey says. Moreover, even if the state does achieve a 35 percent reduction by 2050, that will only put Utah at the level that some of its neighbors already have achieved by 2006, according to URC figures. So the extra 10 percent isn't asking an awful lot.

Washakie and New Water Development Ideas

While the URC and the JVWCD argue over percentages, the elephant still looms in the room. The Bear River Development Act blindly moves us ahead with a life of its own. It requires Utah water managers to devise ways to divert 220,000 acre-feet of Bear River water at the expense of the Great Salt Lake ecosystem. Even if the state has managed to quell public outcry over damming the Bear—diverting water for the Wasatch Front through a pipe into Willard Bay instead—it still has pledged to provide 60,000 acre-feet each to Cache County and the Bear River Water Conservancy District of Box Elder County at some time in the future when these northern Utah folks want their share. That requires a dam.

Right now the most politically feasible site is at Washakie, a depression just west of Portage, Utah. It would be an off-river reservoir behind an earth-filled dam. Water would have to be pumped into it and dikes would have to be built on either side to contain water. Preliminary engineering reports indicate that the dam won't liquefy in an earthquake. The most difficult engineering challenge is rerouting a stretch of the Malad River. However, it will cause the least amount of dislocation to the environment and farming and ranching operations, far less than the Honeyville or Amalga/Barrens sites.

Rancher Fred Selman says the Washakie dam site has some advantages. It would allow the Bear River Canal Company to move water more efficiently to the west side of Box Elder County and possibly make more land available for cultivation, assuming Washakie water would be used for agriculture, which is debatable due to its significant projected costs. Even with a 25 percent subsidy, it would be difficult for farmers to make the figures work, given their tight operating margins.

Moreover, if problems arise with mixing lower-quality Bear River water with water in Willard Bay, the preferred alternative for moving river water to the Wasatch Front, Washakie could supply that water to the Weber Basin and Jordan Valley Water Conservancy Districts. But it would require even more infrastructure costs to convey that water to urban areas because the Washakie site is farther away than Willard Bay.

"In the demand prediction, there are a few isolated places in Box Elder and Cache Counties that could need water by 2050, but there isn't the demand for that 120,000 acre-feet," Merritt Frey says. "I really see this project, at least with the current projections, as a Wasatch Front project. The state seems to agree, too, with the numbers that they've been producing. There just isn't a huge call for northern Utah. The Washakie site and the Willard Bay site aren't as separate as you might think."

Costs are definitely playing a larger role in water development. In May 2005, Utah's Water Delivery Financing Task Force reported that a proposed pipeline from Lake Powell to serve rapid growth in southern Utah, combined with the Bear River project, would cost a combined $500 to 600 million. By July 2005, the same state lawmakers and water officials were saying the total for the two would be at least $790 million.

The reports contained some good news, however. Recognizing that Utah never could afford to build both projects at the same time, the water policy makers said that, due to new water sources and conservation, Bear River water wouldn't be needed until 2025, a full decade later than originally thought. But nested in those reports was a new conflict. The Utah Division of Water Resources and the JVWCD don't agree on how much water we actually have. The state says it's more than the JVWCD believes, even suggesting Bear River water won't be needed until 2030, although it wants to be free of the responsibility of preselling 70 percent of the Bear River project water before construction begins. An August 2005 analysis of community water systems issued by the Division of Water Resources stated that only a few water systems will have a shortfall in 2050, and even then, they will be small shortfalls. Weber Basin is expected to have only a 10,472 acre-feet shortfall, nowhere near the 50,000 acre-feet that the Bear River Development Act would deliver. Salt Lake County would have a 14,693 acre-feet surplus in 2050, although state figures do show that the JVWCD would still need 28,569 acre-feet. The URC looks at the escalating costs and disputed water-supply figures and concludes it's clear that the Bear River project is a waste of money, whatever its cost. It simply isn't needed.

But that doesn't preclude others from making new claims on the Bear, a kind of slow war of attrition. On February 2, 2005, the Federal Energy Regulatory Commission granted the Twin Lakes Canal Company three years to study how it could build an eighty-five-foot-high, seven-hundred-foot-long embankment dam on the Bear River at Oneida Narrows, just below the existing PacifiCorp dam. Immediately eleven groups, companies, or individuals filed opposition to the project: Trout Unlimited, Idaho Rivers United, the state of Idaho, several local residents, and PacifiCorp, which fears it would lose access to its dam and its ability to prevent flooding of the recreation areas it's required to maintain as part of its license agreement. The new reservoir would cover a section of the Bear in Oneida Narrows that the BLM has declared eligible for wild and scenic status because of its "outstandingly remarkable recreational, geologic and wildlife resource values." The new dam would generate only 7 megawatts of power for canal shareholders, besides storing their water rights.

Water Pricing and Property Taxes

Water development in the West largely has been a partnership among federal, state, and local governments and agencies. The Bureau of Reclamation provided the initial resources and built the dams, with individual locales pledging to repay some of those costs with property taxes. Revenues from hydropower generation defrayed some of the costs, but that power, along with water, has been subsidized by taxpayers.

The problem with subsidizing municipal water through property taxes is that it masks the true cost and value of water. In short, it artificially cheapens it. For example, typical Salt Lake County residents pay just less than 50 percent of the cost of the water they use. The rest of the costs are hidden: 19 percent in property taxes, 3 percent in state sales taxes that go into a water-development fund, and 28 percent in federal subsidies, according to the URC.This subsidization encourages waste since residents don't pay out of pocket for all the water they use. Moreover, the property-tax subsidy makes alternative water sources appear more expensive than they really are. The system encourages overconsumption during peak usage, which allows water developers to point to those patterns and figures to reinforce the need for more water development. And as if we needed still another fissure in society, subsidizing water with property taxes has the potential to split communities along a new cultural divide—conservation—because those who conserve water don't see the full economic benefits, and their property taxes wind up subsidizing profligate use of water by those who are environmentally unaware or don't care.

The argument for diminishing the role property taxes play in water pricing is gathering steam. In 2002 the Utah Tax Review Commission recommended that using property taxes for financing water projects be reduced or eliminated. Former Utah Governor Mike Leavitt made the same argument to the state legislature, and Governor Olene Walker added her voice when she took office. The debate is even creating strange bedfellows. In 2001 the URC joined the Utah Taxpayers Association and state Senator Howard Stephenson to promote a bill removing property-tax subsidization for water.

But water developers oppose the change in funding. They want the reliability of the old financing system and worry about the short-term risks of change. David Ovard says his figures conflict with those of people who want change. Moreover, he maintains that property taxes are the only way to get water projects up and running.

"Property tax in the state of Utah—and I think throughout the West—has been a means of startup costs for new districts," Ovard observes. "Without property taxes or some other source of funding, you have no water to sell until the project is complete and delivering water. There's no way to recover revenues. People won't give you bonding. The federal government won't give you a repayment contract."

Economists argue that tiered water rates based on consumption will provide a reliable revenue stream to match the loss of funding from property taxes. Bonds similar to tax-anticipation notes could be sold based on expected revenues from usage, even if conservation cut that usage, once equilibrium had been reached between consumption and a fair water-use rate.

To address Ovard's concern that the JVWCD wouldn't be able to float a bond, Gail Blattenberger, former associate professor of economics at the University of Utah, analyzed the ways other water districts in the West finance their new projects. She found that property tax does not finance water in Albuquerque, Scottsdale, Tempe, Mesa, or Phoenix. These cities utilize water revenue from use, in conjunction with impact and development fees, for financing. They reserve the right to levy a property tax if revenues fall short, but they have yet to do that. The Utah Foundation released a report showing that other western water suppliers are receiving competitive bond ratings, even though they don't levy property taxes. In its own study, the URC found that while all Utah's major water districts rely on property tax, only 22 percent of other western urban water suppliers use that source.

Ovard says that those reports are misleading, if not "totally fabricated." "We got a printout from the Legislature Research Office in California. Out of 864 water districts in California, 47 percent of them are levying a property tax," he states. The discrepancy could lie in the two different samples. The URC looked at water suppliers in all Intermountain West states, Oregon, Washington, California, and Texas. Ovard explains that, while property taxes may not be levied by municipalities retailing water directly to their citizens, they are levied further up the line by water-district wholesalers, who sell that water to the cities.

Ovard also points out that the amount going to water development is just a small percentage of the overall property tax residents pay. He maintains that only about 5 percent of a homeowner's overall property taxes go to water development. But that begs the question. If property taxes are necessary at the front end to finance new water projects, why are they also needed once that infrastructure is in place?

"I can see how you might need property tax to get that bonding for future needs," I offer, "because bond holders would need some kind of reliable way to know that the bond payments are going to be met. You do that with the property tax. But for stuff that's already in place—for the maintenance—is that a separate account that's not paid by property tax, or does property tax go to pay that, too?

"The property tax is used for the water conservancy district. We can use it for whatever we want," Ovard replies. Right now his agency receives eight million dollars in property taxes each year. He needs approximately twenty million dollars a year for capital improvements to support new infrastructure. With a total budget of around fifty million dollars, property taxes only provide 20 percent of his agency's revenues. The rest comes from wholesale and retail sales and impact fees.

When asked what percentage of property taxes are going to develop new water infrastructure, Ovard says that it depends upon what parts of his budget are included. "It's whatever we say," Ovard explains, not trying to be flippant. "If somebody forced us to, I could say that all eight million dollars are needed to meet part of our twenty million dollars of infrastructure needs every year. Some of that infrastructure is for buying water rights; some is for drilling wells and building treatment plants and booster stations and pipelines. Some of it is for repairing aging infrastructure, in replacing pipelines. So I can say people can make a big deal about it, but our annual budget with capital improvements now is fifty million dollars. Our O&M [operating and maintenance] expenses this year in this budget are twenty-four to twenty-five million dollars. Property tax is eight million dollars. You tell me. Do you want us to say that we are putting it toward capital improvements or to O&M? It's not funding either one of them totally. The property taxes do offer a measure of stability in our water rates."

I press Ovard to offer more reasons why the benefits of property taxes going to water outweigh the obvious disadvantages. He mentions commercial properties and fire-insurance rates. "With commercial property, which uses very little water, you have to oversize your mains and your facilities by 20 percent to provide fire protection for them. You may have somebody who has a drinking fountain and a restroom or may have five or ten acres of lumber and buildings, and you are providing the fire protection to reduce their insurance rates. If you take the property tax off the commercial property, and they're paying nothing for the benefit they're getting, you shift the commercial tax burden over to the retail water users."

Ovard then approaches the touchy subject of tax equity and argues that property taxes are more progressive. "Say someone has a million-dollar home," he suggests. "They're paying a lot more in property taxes than someone who has a two-hundred-thousand-dollar home. You eliminate property taxes, there is a tax subsidy there, too. The wealthy are helping to subsidize the residential water for the smaller homes. That also will come on the fixed-income people, on the low-income people, on the small residential residences."

Blattenberger's analysis indicates just the opposite. She says that, traditionally, economics literature argues that property tax is regressive. While property tax is higher for larger homes, poor people spend a greater portion of their income on housing than do higher-income people. Moreover, higher-income people get an indirect federal subsidy because they can take a deduction for property taxes on their itemized returns. Lower-income people, who have less ability to generate itemized deductions, must take the standard deduction and forego writing off a percentage of their property tax that goes to water development.

Besides, media stories during the drought of large homes guzzling water at unconscionable levels simply because the owners can afford the higher rates strains the argument that large homeowners are helping out the little guys with their water needs. Being able to look at the green lawn of a prosperous person probably won't be an equitable emotional trade-off for a middle-income person who has to let his grass go brown during a drought because he can't afford the higher rates.

Ultimately, though, property taxes do not provide incentive pricing, and conservation can't enter the equation establishing the true market value of water. There can be no reward for consuming less unless water pricing is based on use. By not having an economic option of choice, Blattenberger says water users lose what economists call *consumer sovereignty*.

One thing both Ovard and the URC agree on is that, if property tax for water development is phased out and replaced with pricing based solely on consumption, the change will have to be gradual. Ovard needs a stable revenue stream to guarantee bond payments. He worries that water consumption can vary too much and cites the current drought and conservation. If his agency has to rely solely on fees from consumption, and if people wind up conserving a lot of water, he will have a revenue shortfall and may lose his agency's favorable bond rating.

Blattenberger's figures from communities that rely more on water revenue from consumption should offer Ovard some solace. Those numbers show that, over time, equilibrium is reached, and revenues based on water consumption are as stable as property taxes. Blattenberger and the URC also agree with Ovard on another pocketbook issue. People who think that switching from property taxes to water-consumption rates will mean that they pay less for their water are dead wrong. We will all wind up paying more, and probably even proportionately more in the short term, until that equilibrium is reached where conservation and less water use are equalized by higher water rates so that Ovard can have a reliable revenue stream. When that happens, the market will have established the true value of water in a desert.

Water Pricing and Unintended Consequences

Initially there will be consumer shock, and politicians and water-agency bureaucrats will have to take the heat as the water bill grabs a larger percentage of household budgets. But are there other possible unintended consequences of moving to water fees based on consumption?

Ovard and water developers are in the murky business of divining the future. They have to anticipate growth—historically, profligate growth—and deliver the water. If they have a better idea where growth may and may not happen, they can lower their growth projections and need for infrastructure. Enter conservation easements, where landowners pledge not to develop land in exchange for tax advantages. Water managers will probably support more conservation easements because they'll know they don't have to deliver water to that land. So we'll likely end up with more open space.

Water rates will go up faster in the near future than the basic inflation rate. Because water rates eventually will reflect the intrinsic value of the resource, and since that value has been artificially depressed, some catching up will be necessary.

High water users will pay proportionately more for the water they consume. That is already happening, although only to a gentle degree. Some water districts, like Salt Lake City, have tiered rates. All people pay the same basic rate for the first gallons they use, the minimum needed to maintain drinking, washing, and some landscaping, but they pay increasingly higher rates on tiers above that.

New sources of water will become more financially attractive for development, like gray water or converting irrigation water in suburban ditches that still flow even though the farms and orchards they sustained have long disappeared. There will be financial incentives for private enterprise to enter the water industry, probably in partnership with public agencies. For-profit businesses will become water brokers and purveyors. Because water lost to evaporation will have more economic value, conjunctive projects, like pumping surface water from reservoirs into aquifers, will make more sense. That's already happening in Arizona and Cadiz, California, where communities are more attuned to conservation.

Taxpayers and tithers are in for another shock. Right now churches, schools, universities, and other nonprofits like museums, sewer-treatment plants, and water-conservancy districts are exempt from property tax. As a community, our property taxes have been subsidizing water used by these laudable institutions. When these are eliminated and replaced by consumption rates, tithing will have to cover the costs of water used by worshippers, and taxpayers are going to have to pick up more of the tab for the water kids use in schools and colleges, perhaps with increased property taxes levied by school districts. If large users of water, like universities and schools with a lot of green space, don't dramatically cut back their water use, we may find ourselves trying to decide what's more important: textbooks, lower class sizes, or water—for drinking, cooking, landscaping of playgrounds, showers after athletics. Debating the relative worth and cost benefits of these uses won't be pleasant.

Our communities will face new ethical challenges centered on the question: If water is a precious resource, but one that is necessary for individual survival, how much water should each person have? How much water is necessary to live? A parallel way to think about the question is offered by another requirement of life: oxygen. What if oxygen could be controlled and brokered like water? Would we accept the logic of asking how much oxygen is enough? It could be defined as enough to sustain the body for work but not for play or engaging in strenuous activities. Will we find ourselves considering similar possibilities?

Is there an intrinsic right to water? Or should the market and the ability to pay allocate the finite, precious resource? Even if some people can afford to pay for profligate use, is it fair for them to do so if it means less water for others?

Hopefully we will be able to balance the conservative position that the market should establish the value of water with a liberal stance that a purely market-based approach to valuing and allocating water may be efficient but unfair. We likely will have to establish a basic minimum and water-rate subsidies for low-income people. As a community, we will have to affirm our common bond through water. Otherwise, we'll likely see the need for "water banks" similar to food banks to help people during times of acute shortage or hard luck.

Water and Myth

The task of changing our attitudes toward water in the Intermountain West is gargantuan, partly because we must challenge the historical and philosophical legacy of water. Water pulses through our community like blood through a body. It's that fundamental. Yet, as powerful as that bond uniting hydrogen, oxygen, and culture is, we must summon the courage to deal honestly with fundamental problems. We need to begin by jettisoning myths from our past, myths that rationalize our use of water.

One of those myths is manifest destiny. We came to believe we had an inherent right—and obligation—to settle the West. We found the idea appealing that we needed to take control of the land and its resources. The right of a new nation to claim ownership of these resources and treat them like property was promoted. We even transferred them from public to private hands, a process that some would like to accelerate.

Embedded in manifest destiny was the myth that if we were going to exercise that right, we'd need to be fruitful and multiply. We built the myth of the garden and the idea that we needed to make the desert bloom. We took pride, perhaps overweening pride, in accepting the challenge of the land and subjugating it, in scratching a bountiful society from hardscrabble desert. Of course, we needed to extract and exploit the resources to make the desert bloom—and green.

Then we told ourselves that these resources were given to us by God, and He wanted us to use them to prosper and grow. We anointed ourselves as His favored species and announced that we had a divine right to His resources above all other plants and animals. Conquering the land became our mission as we wrapped ourselves in the myth of ecclesiastic determinism.

Those myths may have been essential for the initial development of the West, but they are destructive today, especially when they are publicly assumed to be a priori truth. By falling back on them to support our behavior, we erect a barrier to change. We cannot get to a place of constructive change if we summon and trumpet myths that encourage steadily increasing consumption as the population of the West explodes. We will find ourselves caught in a mythological web of our own making.

Another myth, although one that falls within the shadow of the real, is that we are running out of water. Water developers point to the average fourteen inches of rain at the Salt Lake International Airport, in the flats of our arid high desert, and shout that we face a crisis and need to build one more dam, as if that is the key to our survival. Biologist Alice Lindahl says that chicken-little argument "provides political muscle for dam building."

But the water-supply equation is more complex. The headwaters of the Bear River can collect nearly forty inches a year. The Wasatch Mountains east of Salt Lake and Weber Counties, which would receive 100,000 acre-feet of the diverted Bear, gather even more rainfall, close to forty-eight inches, an amount greater than the annual rainfall of Seattle. Utah's yearly yield is 8.5 million acre-feet, theoretically enough water for thirty-four million people, according to Professor Dan McCool in *Waters of Zion*. With much of the Wasatch Front moisture seeping into aquifers, it would seem that we have enough water for some time to come. Certainly Cache Valley has a lot more accessible groundwater. So we probably need to be thinking about drilling more wells at a cheaper cost than damming and diverting the Bear.

However, if you figure global warming into the equation, it may not be an exaggeration to say we are running out of water. In 2005 we are in the sixth year of drought, even with an above-average water year. It's too soon to know, however, if the local drought is being affected by global warming. We simply don't have the data. We have only now reached almost unanimous consensus that global warming is real and aggravated by greenhouse gasses; only a few know-nothings and political institutions with vested interests are still trying to seed doubt by claiming that global warming is a fiction created by liberal environmentalists. We don't know what the local effects of global warming may be. Scientists in the Department of Meteorology at the University of Utah are mounting a team effort to explore those questions, but they are just starting to *define* the complex dynamics of weather in the Wasatch Range and Bear River Basin, let alone report whether we are likely to have more or less rainfall in the future. Unfortunately, we are on the brink of possibly changing our local weather patterns without having that critical information.

Great Salt Lake and the Lake Effect

The Bear River provides 60 percent of the inflow into Great Salt Lake. If part of that water is diverted to the Wasatch Front or placed behind dams for later irrigation, it obviously won't enter the lake. According to the URC, those planned diversions will take 18 percent of the Bear's annual flow in an average year, 70 percent during the lowest water year. In its 1992 Utah State Water Plan for the Bear River Basin, the Utah Division of Water Resources stated that, of the water currently diverted from the Bear, about 58 percent is lost from the system. In their study of the needs of the marshes of the Bear River delta, John Kadlec and Stephen Adair say that perhaps some of the remaining 42 percent, or 500,000 acre-feet, returns to the river or groundwater over some period of time. But they point out that no one really knows because information on depletion and return flow is difficult to get. "A large part of the uncertainty revolves around groundwater levels and movements, which are not well understood for most of the Bear River Basin," according to Kadlec and Adair. David Ovard is going out on a limb when he says about 50 percent of water diverted for agricultural or municipal uses works its way back into Great Salt Lake. Another way of saying that, is that about 50 percent *does not* work its way back into the lake.

Now consider the dramatic fluctuations of the level of Great Salt Lake. At its historic low of 4,191 feet of elevation in 1963–64, the lake covered 950 square miles. In 1986–87, when it reached its historic peak of 4,212 feet, the lake ballooned to 3,300 square miles, a whopping 247 percent increase in just twenty-three years. At its average 4,200-foot elevation, the lake spreads out over 1,700 square miles. Obviously the surface acreage varies substantially, largely because the lake occupies a shallow depression.

Ironically, we can't even agree on the lake's average depth. Utah.com, the state's official tourism Web site, says the lake averages twenty feet deep. The Ogden Convention and Visitor's Bureau reports a thirteen-foot average depth on its Web site. The American Southwest Web site says the lake now averages thirty feet. The Great Salt Lake Hydrologic Observatory Web site has conflicting information. Its graph of the level of Gilbert Bay reflects the square footages already cited, but the accompanying text states different surface acreages. It says the average depth of the lake is "14 in.," no doubt a typo. Probably the the *Columbia Electronic Encyclopedia* is the most accurate when it says the lake averages from thirteen to twenty-four feet.

Still, that's a fluctuation of eleven feet. It's no wonder, then, that small variations in the lake's elevation result in dramatic changes in surface acreage. These figures also reveal how little we really know about Great Salt Lake. As an example, consider what's come to be known as *the lake effect.*

When the temperature of colder air blowing over warmer water in Great Salt Lake exceeds sixteen degrees centigrade, the front picks up water from the lake and deposits it, usually as snow, on the western slopes of the Wasatch Mountains east of Salt Lake City. These lake-effect bands generate significant amounts of moisture, especially when the northwesterly flow moves the front along the length of the lake and over more water. While meteorologists can't conclude that the large snowfalls are due only to lake effect bands—the process is complex and complicated by orographic lifting, which also helps deposit snow—we do know that the more surface acreage of water the storm fronts pass over, the more water they can lift into the atmosphere and release as moisture along the Wasatch Front.

If the level of Great Salt Lake fell, lake-effect bands couldn't pick up as much water, and there would be less precipitation in the mountains. Since Salt Lake City relies on canyon streams and groundwater for close to 50 percent of its supply, less water deposited in the mountains via the lake effect means less surface runoff and shrinking aquifers—and reduced water supplies. So, if water is diverted from the Bear when it normally would go into Great Salt Lake, it stands to reason that the surface acreage of the lake will fall, causing less precipitation as well as diminished snow depths for a ski industry that is a primary economic engine for Utah. We will be hit with a double whammy. We will lose water going into the lake because of direct diversion, and return flows won't make up the loss. We'll wind up taking water out of the Bear to mitigate drought, yet that very diversion will aggravate the drought by diminishing the lake effect.

This frightening thought acquires real urgency when we remember that the Bear River Development Act decrees that water *will* be diverted from the Bear. We have made this decision without even considering the long-range consequences to our weather patterns and the moisture that must fall in the mountains if we are to survive in the desert. Things may work out, or this may be a cataclysmic mistake. Certainly we should critically examine the dynamic lake effect using good science before we categorically declare we should divert Bear River water.

Diverting water from Great Salt Lake won't just affect humans, either. I asked Al Trout of the Bear River Migratory Bird Refuge how reduced water flow into the lake will impact the margin between the lake and the refuge.

"This whole area up here, Bear River Bay, is a dynamic interface with the fresh flows from the Bear River delta that come down," Trout says. "That moving interface is where all the life happens."

"It's not a comfort for me to know if the lake recedes, we get more mudflats," Trout continues. "Maybe in the south end of the lake, they've felt that over time mudflats have been impacted more by development, and they could use a few more mudflats down there. We've got square miles upon square miles of mudflat. It goes forever. We don't need any more dry mudflats. What we're short on is water supplies to keep some water on the vegetation."

Jack Greene, the retired environmental-science teacher, reminds us that Great Salt Lake is part of the Western Hemisphere Shorebird Preserve. The Bear River is a critical part of that ecosystem. "The Great Salt Lake is one of the hot spots in the world, as far as its value as a nesting and resting place for millions of shorebirds and waterfowl. That lake wouldn't exist except in only a minor, diminutive sort of way if it wasn't for the water that comes in from the Bear River," he observes.

"And still, people want to divert that water," I say.

"And *still,* people would *love* to take water out of the Bear River," Greene replies emphatically. "'Gosh, it's going into a stinky old marsh in this dead lake,'" he says, mimicking an oft-heard complaint. "The Great Salt Lake is one of the miracles of the planet as far as an aquatic system."

Thinking about the effects of diverting Bear River water from Great Salt Lake, I remember the comments of the two water developers at the budget meeting of the JVWCD. They claimed the 1.2 million acre-feet of water going into the lake are "unused" and "wasted." I suppose they are, if you look at the future nearsightedly.

Resolution

Humans have treated rivers with disrespect for some time. Basically we've considered them convenient conduits, carrying away our offal. It's been easier to push the dead tissue of our industry into our rivers and let the current sweep them out of sight, and too often out of mind, than to examine ourselves. We've let rivers become our path of least resistance.

The way we've treated our rivers says much about us. It's been so easy and gratifying to watch detritus drift downstream or seemingly dissolve as the current drags it under. Because rivers were clean when we encountered them, we've assumed they will cleanse us. We've even invented convenient truisms like "a flowing river cleans itself after fifty feet." But we've deluded ourselves, thinking rivers can carry our burden, and in that delusion we've lost control over our appetites and the opportunity to mature. We've lost the chance to dig into our psyches and wrestle with barriers to change. Ironically, we've spurned the gift of rivers. While they can't cleanse us, they can teach us if we listen and allow them to lift us to a new level of understanding.

By conquering rivers, we lost that connection to a nature that once invigorated us, even as it threatened us. We lost the power and energy that come from exercising restraint, from fasting. We lost sight of the dream that we could live inside, rather than outside, nature.

As we consider what we've done and are about to do to the Bear River, we should try to listen to the voice of the river. It's there in the whisper of the current. To hear that voice, though, we need to listen to it from a point outside ourselves. Only when we reach that place can we begin to look inside ourselves.

We need to respect the river and our place in its grand scheme by letting go of the myths that prevent us from seeing that place. We must overcome the tendency to focus on the here and now and envision a longer timeline, one that measures water issues by millennia rather than decades. We need to begin practicing a tough-minded pragmatism.

We can start by looking at water law, first by recognizing that a river is naturally free flowing and leaving water in it is beneficial. While miners needed a reliable source of water to develop their claims in the nineteenth century, the law and customs of "first in time, first in line," "use it or lose it," and "water must be diverted for its use to be beneficial" are outmoded for the twenty-first century. While we have to recognize the property interests tied up in water rights, we need to summon the political courage to unravel the Gordian knot of water law to accommodate an enlightened vision of water management, one that recognizes that there are new, legitimate uses that, while unforeseen in the nineteenth century, may represent a better use of the resource in the twenty-first. Just because it's a new use shouldn't mean that it must be assigned the most junior of all water rights. If it requires buying water rights from willing sellers, we should be willing to tax ourselves and offer a fair price that reflects the true value of water.

We should recognize that using water for agriculture is beneficial because it provides our food, maintains open space, and helps recharge aquifers. We never should find ourselves in a situation where we have to rely on importing all our food from other nations. But we need to look critically at the cost-benefit ratios for raising certain crops. If we get more energy out of some crops that use less water and do less harm to the environment, then we need to move the water to those fields and focus less on pastures.

We ought to make sure we reclaim unfair profits as we convert water from agricultural to municipal and industrial uses. If water was developed by taxpayer subsidy, then the taxpayer should be the beneficiary when the water it subsidized is converted.

Consider the plight of a young couple hoping to own a home one day. They pay taxes for subsidized water for farmers, who then turn around and sell it to residential developers at a price that reflects the higher value of municipal water. When it comes time to buy that house, the young couple can't afford it because the price includes hookup fees for now higher-priced water that they earlier subsidized with their taxes.

There are those who say that "people should come first," that water should be managed for human beings, and if that means accepting a degraded environment, then that price must be paid. They say our economy requires unlimited access to resources. They call themselves *realists* and declare human beings will always take what they need. In effect, though, they are *short-timers* stuck in an Old West, whose attitudes spell doom for any future we may have. If we don't protect our ecosystem and the water that gives it life, eventually there won't be an economy because a sick environment can't sustain one, at least not a robust one and certainly not one like the one we're accustomed to. Any viable economic future relies on a healthy environment, and lip service won't provide the necessary protection.

The problem of water in the West in the twenty-first century dwarfs our energy problem. With energy, we can nervously assure ourselves that, at least, there are alternatives; that one day we'll discover and harness new, clean, abundant, and renewable sources.

There is no substitute for water. What falls from the sky is all we're going to get, and most certainly rain does not follow the plow as we once quaintly believed. Moreover, if increasing numbers of people have to compete for that finite source, there will be less to go around.

Ultimately, we are going to have to wrestle with the problem of population growth in the arid West. Even though birth rates are going down, the in-migration of the twenty-first century poses more of a challenge than the one during the nineteenth century simply because our water resources are finite. Yes, a growing population can accumulate sufficient wealth to afford transbasin diversions of water. But that is only a stopgap measure. At some point, there will be no distant basin to suck dry. If we wait until then, it will be too late.

We need to learn finally to live in nature, rather than trying to overcome, tame, or master it. We need to take the next critical step in the development of the West, leaving behind the emotional and mythical baggage born of exploration, settlement, and manifest destiny of the distant past. We must come to grips with the dark underpinnings of our expansionist past. We need to be willing to give up the crutch of fecundity and the myth of the garden. We need to make sustainable development our cultural grail.

I ask David Ovard if he thinks we can reach a point of sustainable economic development and population stability that will eventually be compatible with our finite water resources.

"My gut feeling would be no," Ovard replies after a moment's thought. "The ideal would be for somebody—and when I say 'somebody,' I don't have a clue who could possibly do this—would be for them to define what was needed for the environment. Then they would define what would be needed long term for agriculture, for open space. Then they would say, 'This is what's left, and this is all you got to work with.' And that would make it a lot easier for us as water purveyors and the general public to know that we are working with a finite amount of water."

"The problem is," Ovard continues, "that we are not mature enough, we are not far enough along yet to define what the water needs are in those other areas. So it's going to be a very messy process. We're just going to muddle along for a number of years before we reach an equilibrium. But at some point, population needs are going to dictate probably a far stricter conservation ethic than we have now."

Ovard then takes a deep breath and offers his take on population growth. "No one wants to be the person to say, 'This is all we can handle.' We don't tell people how many children they can have. That's a basic, fundamental right, and I don't want to suggest that we do that. But if we have all the children we want, therefore we ought to be able to build all the homes we want. We want to have all the commercial activity we need to support that. Until we hit that wall, I don't think we're going to deal with that."

"We're always going to have our children," Ovard continues. "We're always going to have more children than anyone else. But we have a finite resource in water. Water will be the limiting factor for everything eventually."

"So we're just avoiding the big question, then," I respond. "Population growth."

"I think so" Ovard replies. "I've found that I'm always too far out in front of other people. But I think that's the truth. It's just a matter of time. Further down the road, it's going to be more obvious to more people that we have a problem."

Still, Ovard is an optimist. You have to be if you are in the water-development business. He takes some solace in our ability to solve the problem when he looks at the past. "I think you have to go back into the past," Ovard suggests, "to assess what may be possible in the future. I've done that on occasion, and it sometimes helps. If you go back fifty or a hundred years and see how much our culture and society have changed, things that were unchangeable a hundred years ago, we now take for granted."

"We will adapt, and things will be a lot different," he continues. "With regards to water use and water conservation, far more than I can even project in my mind. If the people will just do what's necessary to deal with it."

The 2006 Utah State Legislature ramped up Bear River development. It eased the 70 percent presale requirement by saying that planning and environmental studies can be undertaken before water-delivery contracts are signed. In effect, it said that now is the time to measure the Bear—and by extension, ourselves. The studies could show that we should take the conservative approach, explore less-expensive ways to provide water. Honest studies based on good science will underscore, once and for all, that there are limits and we need to recognize and honor them.

Ultimately, that is the message in the voice of the Bear River. It is telling us that we have time but not all the time in the world. Thankfully delaying the Bear River project until 2025 has given us a bit more time. Time to consider the effects on Great Salt Lake. Time to pledge ourselves to sustainable development, including population growth. Time to shed myths from time past.

How we treat the Bear River will say much about who we are and what we will become. Not only does the river carry into Great Salt Lake what may be our last drop of water. Our future floats on its current, and that current is unyielding. If it dries up, so do our dreams. The Bear River is our last chance to change course.

Selected Bibliography

Bangerter, Arnold. 1967. *Bear River investigations: Cache and Box Elder Counties, 1962–1965*. Salt Lake City: Utah State Division of Fish and Game.

The Bear project: Preliminary design plan for channel stabilization and erosion control of the Bear River, Evanston, Wyoming. June 1992. Boulder, CO: Aquatic and Wetland Consultants, Inc.

Bear River water quality management plan. 1995. Final draft, prepared for Department of Environmental Quality, Division of Water Quality, and Department of Natural Resources, Division of Water Resources. Logan, UT: Ecosystems Research Institute.

Boone, Jim, ed. 1992. *Boating the Bear: An introduction to the Bear River system for users of unpowered watercraft*. Logan, UT: Bridgerland Audubon Society.

Box Elder County comprehensive wetlands management plan. 1999. Final review copy. Salt Lake City: SWCA, Inc. Environmental Consultants.

Christensen, Scott R. 1999. *Sagwitch: Shoshone chieftain, Mormon elder, 1822–1887*. Logan: Utah State University Press.

Clyde, Calvin Geary. 1953. Sediment movement in Bear River, Utah. Master's thesis, University of California, Berkeley.

Crampton, C. Gregory, and Steven K. Madsen. 1975. *The navigational history of Bear River*. Submitted in partial fulfillment of Contract DACWof-75-C0053, Department of Army, Sacramento Corps of Engineers, Sacramento, California, and the University of Utah, Salt Lake City, Contractor.

Dawson, Robert, Peter Goin, and Mary Webb. 2000. *A doubtful river*. Reno: University of Nevada Press.

Eldredge, Scott J., and Fred K. Gowens. 1994. The fur trade in Utah. *In Utah history encyclopedia*, ed. A.K. Powell, pp. 208–13. Salt Lake City: University of Utah Press.

Evans, J.P., and R.Q. Oaks, Jr. 1996. Three – dimensional variations in extensional fault shape and basin form: The Cache Valley Basin, eastern Basin and Range province, United States. *Geologic Society of American Bulletin* 108:1580–93.

Fremont, J.C. 1846. *Report on the exploring expedition to the Rocky Mountains in the year 1842, and to Oregon and North California in the years 1843–44*. U.S. Congress, Senate. London: Wiley and Putnam.

Giller, Paul S., and Bjorn Malmqvist. 1998. *The biology of streams and rivers*. New York: Oxford University Press.

Huchel, Frederick M. 1999. *The history of Box Elder County*. Salt Lake City: Utah State Historical Society and Box Elder County Commission.

Jibson, Wallace N. 1990. *History of the Bear River Compact*. Logan, UT: Bear River Compact Commission.

Kadlec, John A., and Stephen E. Adair. 1994. *Evaluation of water requirements for the marshes of the Bear River delta*. Logan: Utah State University.

Keith, John E. 1995. *The economic impact on Box Elder County of Honeyville dam and reservoir*. A report submitted to Box Elder County Economic Development. Logan: Utah State University.

Lamarra, Vincent. 1997. *Final: a summary of turbidity and total suspended solids: investigations in the Bear Lake Marsh*. Logan, UT: Ecosystems Research Institute.

Leopold, Luna B. 1994. *A view of the river*. Cambridge, MA: Harvard University Press.

Leschin, Michael F. 1997. A hydrogeochemical study of the evolution of the headwaters of the Bear River in the Uinta Mountains, Utah. Master's thesis, Utah State University.

Liljeblad, Sven. 1972. *The Idaho Indians in transition, 1805–1960*. Pocatello: Idaho State University Museum.

Mabey, Don R. 1979. *The bend of Bear River: A history of Caribou County, Idaho before 1860*. Bountiful, UT: Horizon Publishers.

———. 1999. *The lower Bear River—1,000,000 B.C. to 1900 A.D.* Salt Lake City: Atwood and Mabey, Inc.

Madsen, Brigham D. 1983. *The Bannock of Idaho*. Caldwell, ID: Caxton Printers, Ltd.

———. 1985. *The Shoshoni frontier and the Bear River massacre*. Salt Lake City: University of Utah Press.

Madsen, David B., and James F. O'Connell, eds. 1982. *Man and environment in the Great Basin*. Washington, DC: Society of American Archaeology.

McCool, Daniel C., ed. 1995. *Waters of Zion: The politics of water in Utah*. Salt Lake City: University of Utah Press.

McCully, Patrick. 1996. *Silenced rivers: The ecology and politics of large dams*. London: Zed Books.

McIvor, Don, and George Wilson. 1988. *Bear River Bottoms wildlife management area land management plan*. Salt Lake City: Utah Power and Light Co.

Morgan, Dale L. 1947. *The Great Salt Lake*. Albuquerque: University of New Mexico Press.

———. 1953. *Jedediah Smith and the opening of the West*. Lincoln: University of Nebraska Press.

Morrison, R.B. 1965. Predecessors of Great Salt Lake. In *Guidebook to the geology of Utah*, ed. W.L. Stokes, pp. 77–104. no. 20, Salt Lake City: Utah Geologic Society.

Murphy, Robert F., and Yolanda Murphy. 1963. *Shoshone-Bannock subsistence and society*. University of California Anthropological Records, vol. 16. Berkeley: University of California Press.

PacifiCorp. July 1995. *Resource management plan for the Cutler hydroelectric project*. FERC No. 2420. Salt Lake City: Author.

Parson, Robert E. 1996. *A history of Rich County*. Salt Lake City: Utah State Historical Society and Rich County Commission.

Phillips, Nancy. n.d. *Groundwater and surface water: Understanding the interaction: A guide for watershed partnerships*. West Lafayette, IN: CTIC.

Pringle, Laurence. 1985. *Rivers and lakes*. Alexandria, VA: Time-Life Books.

Rich, E.E., ed. 1979. *Ogden's Snake country journals, 1824–26*. London: Hudson's Bay Record Society.

Simmons, Vivian, and Ruth Varley. 1977. *Gems of our valley*. Grace, ID: Grace Literary Club.

Simms, Steven R. et al. 1990. *Archaeological reconnaissance in the lower Bear River marshes, Utah*. Logan: Utah State University.

Smart, Earl. 1958. An ecological study of the bottom fauna of Bear Lake, Idaho and Utah. PhD diss., Utah State University.

Smith, M.E., and M.L. Maderak. 1993. *Geomorphic and hydraulic assessment of the Bear River in and near Evanston, Wyoming*. Cheyenne, WY: U.S. Dept. of the Interior, United States Geological Survey.

Somers, Ray. 1984. *Hampton Ford and stage stop: Collinston, Box Elder County, Utah, Bear River irrigation development*. Logan, UT: Somers Historic Press.

Stauffer, Norman E., and Craig W. Miller. 1992. *Bear River, Bear Lake: Hydrologically, where would they be without being connected?* Logan, UT: Ecosystems Research Institute.

Toth, Richard E. 2001. *Bear River watershed futures study*. Logan: Utah State University Department of Landscape Architecture and Environmental Planning.

U.S. Army Corp of Engineers, Sacramento District. February 1989. *Reconnaissance report: Bear River Basin investigation, Idaho, Utah, Wyoming*. Sacramento: Author.

U.S. Department of Interior, Bureau of Land Management. 1995. *Bear River wild and scenic eligibility: Bear River in Idaho*. Draft resource assessment. Pocatello, ID: BLM, Pocatello Resource Area.

Utah Division of Water Resources. 1976. *Multiobjective interagency study of the Bear River Basin water and related resources*. Salt Lake City: Author.

———. 1991. *The re-evaluation of Bear River reservoir sites*. Final report. Salt Lake City: Author.

———. January 1992. *Utah state water plan: Bear River Basin*. Salt Lake City: Author.

———. 2000. *Bear River development*. Salt Lake City: Author.

———. 2000. *Bear River watershed management unit water quality assessment report*. Salt Lake City: Author.

Vice, Daniel, and Terry Messmer. n.d. *Wetlands of Utah*. Logan: Utah State University Cooperative Extension Service.

Water pollution studies: Lower Bear River Basin. 1956. Cincinnati: Robert A. Taft Sanitary Engineering Center.

Water source investigation of the upper Bear River drainage. 1983. Evanston, WY: Forsgren-Perkins Engineering.

Workman, Gar W. 1963. An ecological study of the Bear Lake littoral zone, Utah-Idaho. PhD diss., Utah State University.

Index

Note: Page numbers in italics refer to illustrations and captions.